THE DIGITAL FUTURE OF MUSEUMS

The Digital Future of Museums: Conversations and Provocations argues that museums today can neither ignore the importance of digital technologies when engaging their communities, nor fail to address the broader social, economic and cultural changes that shape their digital offerings.

Through moderated conversations with respected and influential museum practitioners, thinkers and experts in related fields, this book explores the role of digital technology in contemporary museum practice within Europe, the U.S., Australasia and Asia. It offers provocations and reflections about effective practice that will help prepare today's museums for tomorrow, culminating in a set of competing possible visions for the future of the museum sector.

The Digital Future of Museums is essential reading for museum studies students and those who teach or write about the museum sector. It will also be of interest to those who work in, for, and with museums, as well as practitioners working in galleries, archives and libraries.

Keir Winesmith has been working at the intersection of technology, culture and place for the last 15 years. In 2018 he was featured in *Fast Company*'s "100 Most Creative People in Business" for this work. The former director of digital departments at *San Francisco Museum of Modern Art* and *Museum of Contemporary Art Australia*, he is currently CTO of *Old Ways, New* and Professor at UNSW Art & Design.

Suse Anderson is Assistant Professor, Museum Studies at The George Washington University, and co-producer and co-host of *Museopunks* – the podcast for the progressive museum. A former President of MCN (Museum Computer Network), Anderson teaches courses related to museums and technology, ethics and visitor experience. She holds a PhD from The University of Newcastle, Australia.

THE DIGITAL FUTURE OF MUSEUMS

Conversations and Provocations

Keir Winesmith and Suse Anderson

LONDON AND NEW YORK

First published 2020
by Routledge
2 Park Square, Milton Park, Abingdon, Oxon OX14 4RN

and by Routledge
52 Vanderbilt Avenue, New York, NY 10017

Routledge is an imprint of the Taylor & Francis Group, an informa business

© 2020 Keir Winesmith and Suse Anderson

The right of Keir Winesmith and Suse Anderson to be identified as authors of this work has been asserted by them in accordance with sections 77 and 78 of the Copyright, Designs and Patents Act 1988.

All rights reserved. No part of this book may be reprinted or reproduced or utilised in any form or by any electronic, mechanical, or other means, now known or hereafter invented, including photocopying and recording, or in any information storage or retrieval system, without permission in writing from the publishers.

Trademark notice: Product or corporate names may be trademarks or registered trademarks, and are used only for identification and explanation without intent to infringe.

British Library Cataloguing-in-Publication Data
A catalogue record for this book is available from the British Library

Library of Congress Cataloging-in-Publication Data
A catalog record for this book has been requested

ISBN: 978-1-138-58953-7 (hbk)
ISBN: 978-1-138-58954-4 (pbk)
ISBN: 978-0-429-49157-3 (ebk)

Typeset in Bembo
by Apex CoVantage, LLC

For Scout, Rory, Tali, Ian, and Susan

CONTENTS

List of figures	ix
Acknowledgements	x
Network (im)permanence	xi
Foreword	xiv
Introduction	1
Some provocations on the digital future of museums	10
Conversations on the digital future of museums	**27**
Conversation 1: Seph Rodney + Robert J. Stein	29
Conversation 2: LaToya Devezin + Barbara Makuati-Afitu	43
Conversation 3: Lara Day + David Smith	60
Conversation 4: Sarah Brin + Adriel Luis	74
Conversation 5: Sarah Kenderdine + Merete Sanderhoff	90
Conversation 6: Kati Price + Loic Tallon	105
Conversation 7: Shelley Bernstein + Seb Chan	122
Conversation 8: Kate Livingston + Andrew McIntyre	141

Conversation 9: Brad Dunn + Daryl Karp   158

Conversation 10: Arthur Cohen + Tonya Nelson   172

Conversation 11: Tony Butler + Lori Fogarty   189

Conversation 12: Daniel Glaser + Takashi Kudo   206

Digital practice in museums: where do we go from here?   219

Bibliography   *225*
About the authors   *233*
Index   *235*

FIGURES

1.1	Seph Rodney + Robert J. Stein Conversation Visualisation, Christie Fearns and Keir Winesmith, 2019	29
2.1	LaToya Devezin + Barbara Makuati-Afitu Conversation Visualisation, Christie Fearns and Keir Winesmith, 2019	43
3.1	Lara Day + David Smith Conversation Visualisation, Christie Fearns and Keir Winesmith, 2019	60
4.1	Sarah Brin + Adriel Luis Conversation Visualisation, Christie Fearns and Keir Winesmith, 2019	74
5.1	Sarah Kenderdine + Merete Sanderhoff Conversation Visualisation, Christie Fearns and Keir Winesmith, 2019	90
6.1	Kati Price + Loic Tallon Conversation Visualisation, Christie Fearns and Keir Winesmith, 2019	105
7.1	Shelley Bernstein + Seb Chan Conversation Visualisation, Christie Fearns and Keir Winesmith, 2019	122
8.1	Kate Livingston + Andrew McIntyre Conversation Visualisation, Christie Fearns and Keir Winesmith, 2019	141
9.1	Brad Dunn + Daryl Karp Conversation Visualisation, Christie Fearns and Keir Winesmith, 2019	158
10.1	Arthur Cohen + Tonya Nelson Conversation Visualisation, Christie Fearns and Keir Winesmith, 2019	172
11.1	Tony Butler + Lori Fogarty Conversation Visualisation, Christie Fearns and Keir Winesmith, 2019	189
12.1	Daniel Glaser + Takashi Kudo Conversation Visualisation, Christie Fearns and Keir Winesmith, 2019	206

ACKNOWLEDGEMENTS

This book exists in the world thanks to the guidance and good will of a number of people whom we wish to acknowledge and thank. Firstly, thank you to all the people who were interviewed for this book: Tony Butler, Seb Chan, Arthur Cohen, Lara Day, LaToya Devezin, Brad Dunn, Lori Fogarty, Daniel Glaser, Daryl Karp, Sarah Kenderdine, Takashi Kudo, Kate Livingston, Adriel Luis, Barbara Makuati-Afitu, Andrew McIntyre, Tonya Nelson, Kati Price, Seph Rodney, Merete Sanderhoff, David Smith, Robert J. Stein and Loic Tallon. We also spoke with JiaJia Fei, Ekene Ijeoma, Katrina Sedgwick, and Lanae Spruce for an earlier iteration of this book. Thank you to Courtney Johnston for her foreword.

We must also thank our colleagues and advisors who have provided us with intellectual advice, particularly Rossana Cantu, John Stack, Chad Coerver, Deena Chalabi, Kari Dahlgren and Peter Samis, as well as the reviewers of our book proposal and the manuscript. The editorial team at Routledge, Heidi Lowther, Marc Stratton, and Katie Wakelin helped keep us on track and turn an idea into reality, while Roxann Edwards provided invaluable administrative support.

Finally, it would have been impossible for us to undertake this project without the patience and encouragement our families. Thank you, Susan, Tali and Rory Winesmith, Terry Smith, and Ian and Scout Anderson.

NETWORK (IM)PERMANENCE

We work on, with, and through the internet, and have done so since the 1990s. We are acutely aware that websites, platforms, and services come and go. Let's call it network impermanence.

All web content cited in this book uses the Internet Archive's Way Back Machine to ensure that the exact content being referenced, and its original online context, will be available to readers in the future.

Here's how to read an archived link. The first portion denotes the Internet Archive's web address, the second portion is the time stamp the page was archived onto the Archive's servers and the final portion is the URL that was archived. The below example is a snapshot of the Flickr Commons webpage taken on August 5, 2008.

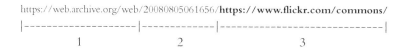

```
https://web.archive.org/web/20080805061656/https://www.flickr.com/commons/
|--------------------|-------------|-----------------------------|
          1                 2                     3
```

If you encounter a web link in the book that does not have the https://web.archive.org/web prefix that means that a specific time stamp for that reference is not necessary. If you encounter any broken links, you can simply enter that link into the Way Back Machine at https://archive.org/web/ and you'll find a snapshot. We've ensured that every page referenced in the book is saved at least once in the Way Back Machine, and we encourage you to do the same with web content that you cite in printed or digital material.

We both use social media platforms for research and collaboration, we've developed social media strategies for museums, and Suse has used social media in her teaching in the Museum Studies program at The George Washington University.

For this book, we encourage readers to use *#DFMBook* for the book as a whole and *#DFMBook1* for specific discussions (in this case *Conversation 1: Seph Rodney + Robert J. Stein*).

We realise that our readers will use whatever platforms they choose and whatever tags (or lack thereof) that appeal to them. If you'd like your thoughts and criticisms to be found by those who don't already follow you, or to appear outside the platform bubble you've been algorithmically assigned, we encourage you to use these tags.

For the front cover we approached the San Francisco based visualisation and analytics design firm Stamen Design, with whom Keir had worked in the past. Using publicly available collection APIs from the *San Francisco Museum of Modern Art* and *M+*, Keir created a dataset that included artist birth and death years, and all artworks by that artist in each institution's collection. It also includes when the works were acquired by the two institutions. This data was supplied to Stamen who used it to create a series of visualisations that show acquisitions across artists' lifespans, expressing how artworks are collected in life (pink) and in death (blue). The blue dots include a blur effect intended to convey an otherworldly atmosphere. From this series we selected the cover image.

"When we make space for others, we make space for possibilities beyond our imagining."

– *Nina Simon*[1]

---

[1] Simon, Nina. 2019. "From Risk-taker to Spacemaker: Reflections on Leading Change." Museum 2.0. 2019.

FOREWORD

Courtney Johnston

The distinguishing characteristic of the people leading practice in the development of "digital" in museums is, I believe, that they are given to both action *and* theory, with a healthy side in activism. Many other industries and occupations combine a strong theoretical aspect with hands-on practice, but I see something unique in the way people who work on the web create, theorise and connect, perhaps because the mechanism we connect *through* is also the stuff we create *with*.

The Digital Future of Museums is, for me, a book about thinking. It's a book that brings into conversation a group of thinkers who characteristically do their thinking in public. The body of knowledge captured in these conversations has grown largely through the mechanisms of conferences, blogs and social media; indeed, it is through these channels that Keir Winesmith and Suse Anderson first encountered many of the contributors. As Seb Chan observes in his conversation with Shelley Bernstein, when Keir asks why both have such a commitment to working "in the open":

> For me it's about working in public. I grew up with the web, and this notion of "viewing source code" has infused my philosophy around lots of things. That we all get better if we share information, and we bring the other generations along with us. And it is about transparency and openness as a differentiator, too. I found that early on at Powerhouse. It was a way of creating an international network of peers and peer institutions that we could bounce ideas off and share information with.

Research, experimentation, documentation and reflection: by talking not only about *what* they are doing, but *how* and *why* they are doing it, and what they have *learned* from doing it – at an unprecedented volume, unhampered by the publishing schedules or peer review of academic journals – these thinkers have shaped the discourse around museums in a new, powerful and rapid way. Moreover, this discourse has been shaped by interactivity from the outset: blogs were made to be commented on and hyperlinked, wikis to be edited, social media posts to be shared and replied to. This is one of the reasons the interview format adopted in *The Digital Future of Museums* serves its contributors and its readers so well: it supports the collaborative development of ideas fostered by the best online engagements, as well as humour, respectful disagreement and new insights.

This ethos of thinking in public is fuelled by tremendous generosity, another characteristic of this community. It is also an approach that takes advantage of the democratising effects of the internet. In this new paradigm it's not the brand name of the institution you're affiliated to but the quality of your contribution that matters. As Seb also observes, this has been particularly important for those of us working in or alongside cultural institutions beyond the traditional centres of power and attention. In the first wave of museum/digital practice, it was the voices of those people working outside the dominant centres of geographical and institutional power that had new power to be heard. Today, it's voices from beyond the dominant cultures – communities that have traditionally been marginalised, silenced or spoken *for* by the museum – that are at the forefront of changing museum practice.

The Digital Future of Museums is also a book about museums as generators and conduits of thinking. Seph Rodney identifies in his conversation with Robert Stein that museums play an important role in "having us think about our thinking, *publicly*." Museums reflect *and* shape the perspectives of nations, societies, communities and individuals about their histories, their cultures, politics, biases and view of the future. By driving a museum's intellectual agenda, funders, management, and staff create the stories and imagery that inform (or confirm, or challenge) how a society thinks about itself. This is why museums are not, and can never be, neutral. It is also why they are such exciting and meaningful – and occasionally hazardous – places to work.

I first encountered Seph's writing in 2016, in a piece he wrote for *Hyperallergic*, based on his PhD research (Rodney 2016). In the essay, Seph traces a progression in museums from the late 19th-century paradigm of the specialist collecting institution, to the post-Second World War paradigm of the educational institution, to the 21st-century paradigm of the visitor-centred museum.[1] In the contemporary context, personalisation has succeeded education. The visitor is no longer seen by the museum as an incidental beneficiary of its core focus on collecting and categorisation, no longer viewed as a passive recipient of authoritative

---

1 For a fuller explication of these ideas, see Seph Rodney. 2015. "How Museum Visitors Became Consumers." CultureCom. 2015. https://web.archive.org/web/20190712193442/**https://culture-communication.fr/en/how-museum-visitors-became-consumers/**

curatorial knowledge, but invited to be an active participant, making their own meaning from the museum's offerings.

Seph attributes the change over the past 25 years to an intertwined set of economic, social, political and museological changes. The emergence of the new museology has placed the visitor at the centre of museum practice; cultural policies (largely in countries where museums are publicly funded) have pushed museums towards new success measures based around access and economic contribution; and the rise of the experience economy has led to the creation of museum experiences designed for consumption, according to the user/visitor's needs and desires. This framework provides a thought-provoking context in which to locate the 12 discussions that make up *The Digital Future of Museums*. The concerns of people leading digital change in our museums far exceed questions of what file format to use, or whether it's AR or VR we should bank on for visitor appeal. Within these conversations, we can see leading thinkers analysing in real-time some of the greatest limitations and opportunities that museums as social structures and services are working through today.

Throughout the conversations, you'll notice contributors offering different versions of a similar refrain: digital isn't a technology, it's a way of working; digital isn't a technology, it's a language; digital isn't special, it's just what is. This refrain brings us to the heart of this book: the fruitful, fascinating and fast-paced ways in which the development of contemporary digital technology has interfaced with contemporary museum thinking.

Like the earliest iteration of the museum, the internet was initially a research tool used by specialists and those with privileged access to (computing) power. When Tim Berners-Lee began development of the World Wide Web in 1989, his goal was to make the internet "a collaborative medium, a place where we can all meet and read and write" (Carvin 2005). With the launch of the Mosaic web browser in 1993, the internet went mainstream: now, anyone with an internet connection could "surf" this massive new global resource for information and entertainment. At this point, however, publishing power was still limited: the means of creating and consuming content were separated. The web, in some ways, was in the second museum paradigm: a place where those with access to the tools of publishing fed content and knowledge to those who primarily consumed without contributing back. We can map the beginnings of digitised collections to this moment, and the new ability for all museums to think massively beyond their local remit into global reach.

It was at the turn of the 21st century, with the rise of blogging and wiki platforms and the integration of publishing tools into the medium, that Berners-Lee's vision was fulfilled: the web as a two-way system, where "ordinary" people could contribute content as easily as they accessed it. This was the read/write web: a medium capable of being displayed (read) and also modified (write). The congruence here between the new affordances of the web and the new approaches to exhibition and experience design fascinate me; all fed by people's growing expectations that their experiences and opinions warrant recognition and sharing.

In 2005, Tim O'Reilly (2005) published *What is Web 2.0*, an analysis of design patterns and business models that he saw underlying the companies that had survived the 2001 dot-com bubble burst. The codification of Web 2.0 highlighted collaborative services, "harnessing the collective intelligence," and the power of user-generated content. This was the age of Wikipedia and Flickr; of folksonomies and tagging; of RSS and viral marketing. It's when *Time* magazine made its 2006 Person of the Year *You*, recognising the millions of individuals who were now populating the web's content, and when ideas of co-creation, co-curation and community participation went mainstream in museums.

Writing in 2017, technology commentator Alexis Madrigal observed that "Web 2.0 was not just a temporal description, but an ethos." It was a picture of the truly open web; an almost Utopian vision of humankind working together for global betterment. "But then," writes Madrigal,

> in June of 2007, the iPhone came out. Thirteen months later, Apple's App Store debuted. Suddenly, the most expedient and enjoyable way to do something was often tapping an individual icon on a screen. As smartphones took off, the amount of time that people spent on the truly open web began to dwindle. . . . The great cathedrals of that time, nearly impossible projects like Wikipedia that worked and worked well, have all stagnated.
>
> *(Madrigal 2017)*

Madrigal's analysis matches with a feeling I developed in 2015, as I toured around museums in seven American states studying digitally enhanced visitor experiences (Johnston 2016). Overall, as I visited the museums leading in this space, I observed a move away from projects focused on global audiences, and a move towards personalised experiences delivered via smartphones and customised devices available on-site at the museum. While these new experiences were often rich, subtle and playful – and designed to collect information that helped institution learn more about the behaviour and motivations of physical visitors – this shift in focus from global to local seemed to be paired with a reduction in sector-wide, collaborative endeavours. I asked myself at the time if this was a natural consequence of digital leaders' roving curiosity: that the opportunities of collaborative platforms, APIs, metadata sharing and so on had been thoroughly explored, and new ground opened up for exploration. I also quietly questioned whether museum leadership and external funders had abandoned the altruism encouraged by the open web environment, and were now more inclined to support projects where the benefits could be firmly located within the museum's walls.

Today, the ground has shifted again. As Adriel Luis observes in his conversation with Sarah Brin, over this intervening period "museum conferences everywhere have gravitated towards the theme of social change."[2] The sector is asking itself hard questions about impact, social change and social justice. Within this questioning, there is a growing awareness that in their efforts to decolonise themselves – to cast off their colonialist and imperialist origins – museums may fall into perpetuating a form of neo-colonialism, often in the attempt to fulfil the vision of universal knowledge they were founded upon.

Sarah Kenderdine touches on the tension between the value museums place on access to heritage and knowledge and the growing understanding of the prejudice of this attitude in her conversation with Merete Sanderhoff. Talking about the creation of high-resolution digital facsimiles of heritage objects, she notes:

> Critics look at this as a kind of massive cultural appropriation or neo-colonialism that the digital has unleashed in the documentation of other cultures, under a banner of heritage at risk. It's an extraordinary moment. It's not like iconoclasm is new, but now it's in the public domain. We are able to create phenomenal amounts of digital data, but it does come with this overtone of neo-colonialism.[3]

The same point is raised in a response signed by over 100 scholars and practitioners in intellectual property law and material and digital cultural heritage to the 2018 Sarr-Savoy Report, which recommended to the French government the restitution of African material culture from French museums to colonised countries. The response challenges the report's recommendation that all items identified for return should be systematically digitised and made available online, arguing that "the report's recommendations, if followed, risk placing the French government in a position of returning Africa's Material Cultural Heritage while retaining control over the generation, presentation, and stewardship of Africa's digital cultural heritage for decades to come" (Pavis and Wallace 2019).

As we look to the future then, I recommend one particular chapter in this book to all readers. The conversation between LaToya Devezin and Barbara Makuati-Afitu shows, in applied practice, ways of working that could and should form the culture of museums of the future. LaToya, in her discussion of working with African American and other communities, describes a post-custodial model of museum and archival practice, where emphasis is placed not on the physical housing of the object, but the preservation of a community's culture and stories – of the institution being of service. Barbara, when asked what is required for institutions to respect and protect the stories and knowledge communities chose to share with them, replies, "It's someone fully understanding culturally and spiritually the sacredness of that knowledge that isn't ours, that isn't mine, and wanting to know how we can help – using different platforms – to ensure

---

2 Adriel Luis, in Conversation 4
3 Sarah Kenderdine, in Conversation 5.

its safety." And when asked at the conclusion of the conversation what museums may look like in 20 years' time if they follow these practices, LaToya concludes:

> We're going to have a more multidimensional viewpoint of history, and I think we can really affect social change. We'll have more diversity within our staffs, and a more enriching experience as a whole. I feel like our space will be more transformative over time, and that we can grow with each other, and maybe it will give us the space to have some of those difficult conversations to be able to foster change. If we all could participate in conversations like the conversation that we're having now, I can see it being such a beautiful and wonderful process.[4]

The beauty of this book is being invited into these conversations in this moment. On behalf of all its readers in the future, I give my thanks to the generosity of all those who gave of their time and knowledge to create it for us.

References

Carvin, Andy. 2005. "Tim Berners-Lee: Weaving a Semantic Web." A Sense of Place Network. 2005. http://web.archive.org/web/20100606132725/**http://www.cbpp.uaa.alaska.edu/afef/weaving%20the%20web-tim_bernerslee.htm**

Johnston, Courtney. 2016. "Final Report: Visitor Experience in American Art Museums." Best of 3. 2016. http://web.archive.org/web/20181001033400/**http://best-of-3.blogspot.com/2016/06/visitor-experience-american-art-museums23.html**

Madrigal, Alexis C. 2017. "The Weird Thing About Today's Internet." *The Atlantic*. 2017. http://web.archive.org/web/20170520025036/**https://www.theatlantic.com/technology/archive/2017/05/a-very-brief-history-of-the-last-10-years-in-technology/526767/**

O'Reilly, Tim. 2005. "What Is Web 2.0? Design Patterns and Business Models for the Next Generation of Software." OReilly.Com. http://web.archive.org/web/20100508031108/**http://oreilly.com/pub/a/web2/archive/what-is-web-20.html**

Pavis, Mathilde, and Andrea Wallace. 2019. "Response to the 2018 Sarr-Savoy Report: Statement on Intellectual Property Rights and Open Access Relevant to the Digitization and Restitution of African Cultural Heritage and Associated Materials," March. http://web.archive.org/web/20190610002217/**https://zenodo.org/record/2620597**

Rodney, Seph. 2015. "How Museum Visitors Became Consumers." CultureCom. 2015. https://web.archive.org/web/20190712193442/**https://culture-communication.fr/en/how-museum-visitors-became-consumers/**.

———. 2016. "The Evolution of the Museum Visit, from Privilege to Personalized Experience." Hyperallergic. 2016. http://web.archive.org/web/20180629102144/**https://hyperallergic.com/267096/the-evolution-of-the-museum-visit-from-privilege-to-personalized-experience/**

---

4 LaToya Devezin, in Conversation 2.

INTRODUCTION

Within the museum technology space, conferences, along with blogs, social media and more formal publications, have played a critical role in the development of a discourse on digital[1] and its meaning for cultural institutions. Indeed, we have been attending, presenting at, moderating, programming and keynoting conferences, symposia and workshops for over 15 years.[2] We continue to do this work, finding great value in it. Most of these events explore contemporary practice with a focus on the intersection of the GLAM[3] sector (galleries, libraries, archives and museums), culture and technology. During these events, in our experience, the best moments often occur in the margins and outside the formal sessions – in hallways, conference hotel lobbies, cafes and bars when a small group forms to actively debate crucial issues that impact our sector. In these dialogues we've encountered many transformative ideas. Many of these discussions,

1 In this book the authors and participants use the term digital in a broad sense to encompass digitally mediated technologies, platforms and services (apps, websites, smartphones, computer networks, public and hosted platforms, collection management systems, etc.), digitally informed practices (including iterative, user tested, speculative and experimental ways of working) and the contemporary, networked, technologically mediated world of the Anthropocene where digital technologies and practices shape society through their role in, and impact on, social, political, economic, cultural and other systems.

2 These conferences include: Museum Computer Network (MCN), Museums and the Web, the American Alliance of Museums conference, National Digital Forum in New Zealand, SXSW in Austin, Museums & Galleries Australia conference, Museums Computer Group in the U.K., MuseumNext, Communicating the Arts, REMIX, AIGA and many more. Suse was Co-Chair of MCN 2015-16, and on the Program Committee of MCN and the Museums Computer Group for several years.

3 We are focused on museum making in this text, however, we have found inspiration in the wider GLAM/GLAMAZONS sector (GLAMAZONS is galleries, libraries, archives, museums, aquariums, zoos, observatories and nature spaces). We have edited the book to create inroads and pathways for those working in, with, or studying institutions other than museums.

along with more formal moments in panels, workshops or break-out sessions, have opened up our thinking, created a new sense of what's possible, encouraged us to try different things, or (most commonly) warned us off strategies or paths that are filled with pitfalls, saving us valuable time and resources. The honest reflections of a peer can provide a solution to a seemingly intractable internal problem, a transformative story of failure – which are rarely presented on stage – helps you avoid making the same mistake, or a single comment encourages you to try something you'd previously thought impossible or unproductive.

From these exciting generative moments (listening, asking questions or offering ideas) a pattern emerged. The best of these discussions are with groups of two to four people that include experienced domain experts from different backgrounds and different museological contexts, people who employ contrasting approaches or strategies to address shared problems. The very best of these uncover new knowledge in the spaces in-between. Daryl Karp notes, "a lot of the exciting ideas come out of informal conversations where there is no real agenda."[4] It was conversations such as these that ignited the initial spark for this book.

These experiences encouraged us to develop a conversational model for this publication. Consequently, and intentionally, the heart of this book is a series of twelve facilitated conversations between paired thinkers, makers, doers and leaders which explore key issues that contextualise the digital present and inform the post-digital future of museums. Grappling with important, contemporary questions through conversation – rather than in a formal paper or essay – generates different types of insights, in a manner that is rare in our sector. Of course, interviews of this nature of inherently impartial. It is worth acknowledging that we did not seek to minimise or prevent bias in our questions, nor did we verify the participants' statements with their colleagues, or users of the projects they cite, to confirm their accounts. Rather than a summary of recent and successful projects, the participants speak from positions of personal and lived experience in order to provide the reader with the context and strategy that underpins their practice more broadly. It's crucial to the success of this book that the ideas are legible and free of museum or technology jargon, because, as Andrew McIntyre says, "you can't change the world unless you bring people with you."[5]

This book takes a non-traditional approach to research. Rather than seeking to follow a single line of questioning, the book's diverse cohort explore a range of distinct, yet interconnected, topics. These topics were developed by the authors reflecting the museum trends we had identified, and include critical input from colleagues, peers and the participants themselves. Each conversation, recorded in real-time, focuses on a specific set of concerns, with room to consider relevant ideas, practical experiences and specific projects. Taken together they provide context for a future which, according to Seb Chan, will be populated by those for whom "digital (and technology) isn't special, it just is what is."[6]

4 Conversation 9: Brad Dunn + Daryl Karp.
5 Conversation 8: Kate Livingston + Andrew McIntyre.
6 Conversation 7: Shelley Bernstein + Seb Chan.

The conversation with the writer and academic **Seph Rodney** (Jamaica/U.S.) and established museum technology leader **Robert J. Stein** (U.S.) begins with a discussion on personalisation of the museum visit and museum content. That conversation broadens out to explore and problematise some different modes or models of museum making from the recent past, present and possible future. Seph and Robert were brought together as they have, respectively, extensive published research and high-profile public-facing project experience regarding the customisation and personalisation of the museum visitor experience.

In 2016 Keir heard the community archivist **LaToya Devezin** (U.S.) in conversation at the Cultural Heritage & Social Change Summit in New Orleans, U.S. talking about her holistic, community-centred approach to digital and non-digital archiving within communities undergoing change or stress. In 2017 the community engagement facilitator **Barbara Makuati-Afitu** (New Zealand) spoke at the National Digital Forum in Wellington, New Zealand about the Pacific Collections Access Project and the community-centred approach *Tāmaki Paenga Hira Auckland War Memorial Museum* was taking with 13 Moana Pacific Island nations and groupings. By bringing LaToya and Barbara together in conversation we were able to investigate the parallels and distinctions between their respective internal and external strategies for archiving, enriching and opening up cultural collections, artifacts and their stories.

Hong Kong's *M+* and *Asia Art Archive* (AAA) are very different organisations in terms of scale, collection, audience, remit and history. However, they are both relatively young multilingual and multi-modal organisations situated in a highly mediated society. We have been following both organisations attentively for several years, the AAA in particular is frequently referenced by peers working in collections and registration departments, or in archives and libraries. **Lara Day** (Hong Kong) and **David Smith** (New Zealand/Hong Kong) offer a unique perspective on how to balance local relevance with a continental remit, in ways that were (at the time of the recording) expressed primarily online.

Both **Sarah Brin** (U.S./Denmark) and **Adriel Luis** (U.S.) joined the museum world via circuitous paths and work in progressive artist-, visitor- and community-centred ways. Sarah's experience working with gaming and technology-focused pop-ups and ephemeral curatorial projects in the museum and commercial sectors led to her participation in the Horizon 2020-funded GIFT project that explores hybrid forms of virtual museum experiences as a researcher and development manager at *IT-Universitetet i København* in Denmark. Adriel, a self-taught musician, poet, curator, coder and visual artist, developed the innovative *Culture Lab* series with his colleagues at the *Smithsonian Asian Pacific American Center*, a museum without a building. We paired them to discuss agency, expertise, play, trust and institutional change, both inside the traditional museum context and beyond, contrasting American, European and Asian perspectives where possible.

Both **Sarah Kenderdine** (Australia/Switzerland) and **Merete Sanderhoff** (Denmark) are based in Europe working on and within museums, as well as playing leadership roles in pan-institutional organisations such as Europeana and the Australasian Association for Digital Humanities. At the time of this conversation, they both held unique job titles; Sarah as the Professor of Digital Museology at *École Polytechnique Fédérale de Lausanne* in Switzerland and Merete as the Curator of Digital Museum Practice at the *National Gallery of Denmark*. The stories behind these job titles begins a conversation that covers concepts such as openness, data sovereignty, ephemeral culture, participation, empowerment and risk – and the technologies that are used to facilitate or express these concepts such as data APIs, data translation, virtualisation, visualisation, simulation and immersion.

Loic Tallon (U.K./U.S.) has been a central figure in the museum technology space since he began organising an online conference series on mobile technology in museums in 2009. He later joined the *Metropolitan Museum of Art* in New York as a mobile manager and rose through the ranks to become their Chief Digital Officer. While at the Met, Loic published a series of articles on the Met's digital structure and future. At about the same time, **Kati Price** (U.K.) of the *Victoria & Albert Museum* in London published and presented a fascinating survey on how museums resource, fund, and structure their digital work and their digital departments. Producing and sharing this sector analysis and insight built on Kati's long track record of publishing the ideas and practices behind the V&A's own digital initiatives. Kati and Loic's discussion explores digital practice and team structure at scale based on their experiences working with large and diverse collections at two of the most visited and photographed museums in the world.

Shelley Bernstein (U.S.) and **Seb Chan** (Australia) have been active in the museum technology space for more than 10 years, rising to senior positions within their respective institutions, and were both in Chief Experience Officer roles when their conversation was recorded. Shelley and Seb are visible members of this community through their publishing and speaking engagements, and for the highly publicised digital projects they've led. In late 2016 they engaged in an asynchronous discussion by commenting on each other's posts on the online writing platform Medium, having appeared at conferences together in the past. This short comment thread, with its open and generous back and forth, suggested a real-time conversation would enable a rich and nuanced discussion. This wide-ranging chapter considers data, privacy, relevance, resilience, collaboration, brand and cultivating new audiences, and compares custom, consortial and third-party digital platforms.

Kate Livingston (U.S.) and **Andrew McIntyre** (U.K.) both founded and continue to lead businesses that focus on visitor experience evaluation and insights. We intentionally began the series with a deep discussion that centred our thinking on the needs and motivations of current and potential museum audiences. Kate and Andrew were selected as they were two of the most thoughtful and nuanced practitioners we'd encountered working in this space. Keir, during his time at *SFMOMA*, worked with both Kate and Andrew directly evaluating digital projects. This familiarity made possible a discussion that includes a mix of broad theoretical and detailed methodological moments.

We brought **Brad Dunn** (U.S.) and **Daryl Karp** (Australia) together to discuss how cultural organisations can present as media companies, employing broadcasting or journalistic principles and strategies, and how these practices can affect how people experience a museum visit, how they engage in person and online, and the trust they place in museum content and institutions themselves. Brad and Daryl bring their prior experience working in media to their museum work in different ways. Brad, as Web and Digital Engagement Director for the *Field Museum* in Chicago, brings his media, theatre and game production experience, and a journalism practice to the *Field*. Daryl, as Director of the *Museum of Australian Democracy* in Canberra, employs large- and small-scale media content production models at the museum, based on her experience in leadership roles within the Australian Broadcasting Corporation and Film Australia.

The 2017 iteration of the long-running "Culture Track" report from LaPlaca Cohen and the 2018 "Culture is Digital" report from U.K.'s Department for Digital, Culture, Media & Sport are two important data points in the longitudinal view of cultural definitions, access, visitor engagement, and analytics for the cultural sector in the U.S. and the U.K. We spoke to LaPlaca Cohen co-founder **Arthur Cohen** (U.S.), a leading strategic thinker and consultant in the cultural sector, especially for museums and galleries, and **Tonya Nelson** (U.S./U.K.), who was Head of Museums and Collections at University College London, and one of the authors of the "Culture is Digital" report, about shifting definitions of culture, changing audience expectations and motivations, points of access and inclusion, and new models of funding and philanthropy in the museum sector.

Tony Butler (U.K.), Director of Derby Museums Trust which includes the *Museum of Making* (opening 2020) in Derby, U.K. and **Lori Fogarty** (U.S.), Director and CEO of the *Oakland Museum of California* (OMCA), in Oakland, U.S. are directors of two hyper-local organisations that act as exemplars of progressive, audience-centred museum practice. Both Tony and Lori, and the organisations they led, have used audience evaluation and understanding to inform their exhibition and public programs, as well as the supporting organisational structures. In different ways, they both seek to engage all audiences from those who are simply visiting the building to those who participate in physical and conceptual co-production of museum content and experiences. They, and their respective teams, have placed social impact, wellness, trust and genuine relationship building with community at the heart of their very contemporary organisations.

Daniel Glaser (U.K.) is a self-described "brain person" with a diverse experience working at the intersection of science and arts, including as the founding director of *Science Gallery London* at *King's College*. Well known as a neuroscientist for his column in *The Guardian* newspaper, Daniel was the world's first scientist-in-residence at an arts institution at the *Institute of Contemporary Arts* in London, U.K., in 2002. **Takashi Kudo** (Japan) is an artist, designer and communications director with teamLab in Tokyo, Japan. teamLab are an artist collective who create large scale immersive installations and exhibitions that probe technology, nature, and human experience. We brought Daniel and Takashi together to examine participation, play, inclusion, authorship, technology, and social impact at the intersection of science and art in museums.

Selecting participants in order to present a diverse range of experiences, opinions, institutional knowledge and local contexts is critical to the efficacy, structure and broad relevance of this publication. The interview cohort is half female and half male, located across Europe, the United States, Australia, New Zealand, Hong Kong and Japan. The majority of the cohort are museum insiders, with many in leadership positions including director, CEO, chief digital officer, chief experience officer, web & digital engagement director, head of collections & digital experience, head of digital media, curator of digital & emerging practice, and curator of digital museum practice, with others in mid-level roles, such as community archivist, community engagement facilitator and manager digital programme. Approximately a third of the participants are not museum professionals, instead working with or researching museums as an artist, designer, writer, evaluator, strategic consultant or university educator.

Takashi Kudo uses the metaphor of territories when talking about *teamLab's* creative process, "if we try to solve something, it's from within our territories."[7] Each of the participants approach the questions and topics in their conversation from their territory. Some are not digital specialists, they do not have "digital," "technology" or "media" in their job titles. They are, however, engaged in understanding how digital is changing museum work and have experienced digital's effect on their organisation and their working practices.

On a practical level, we solicited and confirmed participation in-person or over the phone months before the recordings. A number of people agreed to take part that were not included in the book after we completed our research and made the final pairings. Once the pairings were confirmed, further research was undertaken, topics were selected, questions and prompts were written and the recordings were scheduled. Lastly, we provided participants with a list of high-level topics for discussion, but not the questions themselves, a week before the recording. This ensured the ideas were front-of-mind, but their answers were not rehearsed. The conversations themselves lasted for between 90 and 120 minutes.

Transcripts of the conversations went through an extensive editing process, wherein Suse tackled several initial rounds of edits, with each interview edited down significantly. Edits were made for readability and to select for ideas and themes that connect across the publication, with a mind to ensuring that no important context or content was lost, and that the unique voice of the discussant was not overridden. Following this initial editing process, Keir would review the written document, and reference the taped discussion when useful, to ensure that the edited discussion still reflected the tenor of the live conversation. All participants were then sent both an edited and unedited copy for review to ensure

7 Takashi Kudo, in Conversation 12.

their edited transcript was accurate and legible, and reflected their views and experiences. Most of the transcripts include a number of short breakout texts in the footnotes that provide background information on projects or processes the participants reference. This detailed preparation, recording and editing process enables the conversations to go beyond the expected and into new territories, offering new knowledge, new connections, and rich, thoughtful context for the digital world that museums increasingly inhabit.

This publication does not seek to be all-inclusive, nor can it be. As with any potential grouping, there are gaps. We cannot chronicle all voices or expertise; however, we can offer a diverse, fascinating, interconnected set of ideas, examples, and memorable moments of creative and productive tension. Nor are we trying to cover all topics of importance to museums today. Instead, we explore specific territories of interest for deeper insight. Although we sought to include voices from many different museum contexts, we primarily spoke to those who speak English as their first language, and did not explicitly connect with people from Africa, South America or the Middle East. This, in part, was due to the specific topics we were interested in covering, and in part was due to our own greater familiarity with the sectors in the English-speaking world. We must also acknowledge our own biases; both authors are white, cisgender Australians who lived and worked in the United States for the majority of this project. For each conversation, Keir wrote the questions and facilitated the discussion over the phone or via video conference. Suse reviewed the questions, with support from Rossana Cantu and John Stack who reviewed questions when they overlapped with their expertise, however all conversational cul de sacs are of Keir's making.

All the interviews were recorded over a 12-month period from mid-2017 to mid-2018, frequently across three time zones and occasionally in the middle of night San Francisco time. The essays in the book, including the introduction and the foreword, were all written in 2019, informed by the conversations herein, but also shaped by the broad technological and cultural shifts happening within and beyond the museum sector. It is therefore important to understand this publication as providing both a snapshot in time, and a timeless, reusable, reference for digital museum practice. Where possible we have tried to steer these conversations away from the trending digital gadgets of the day, augmented reality (AR), virtual reality (VR) and artificial intelligence (AI) being foremost among those in 2017 and 2018, as these will inevitably fade as newer, shiner distractions appear. We look, instead, at how digital impacts all museum work in organisations that are, as Lori Fogarty says, "trying to move away from permanent collection galleries or permanent exhibits to building and prototyping and iterating."[8]

This book does not provide a list of recently completed digital case studies, this material can be found in the conferences and their proceedings referenced earlier in this chapter. We have not asked participants to write an essay, instead

---

8 Lori Fogarty, in Conversation 11.

we have asked them to reflect, in an effort to essentialise and synthesise, not simply to document. In her conversation with Brad Dunn, Daryl Karp says "digital for me is a language. It is not just a technology."[9] This book provides texture and context for the language of digital as it evolves around and within museums.

Perhaps the most consistent thread that ties each of these discussions together is a focus on relationships – relationships with diverse communities, relationships with "big tech," collaborative relationships with other institutions, internal stakeholder and institutional relationships, relationships between the different conversations and the practitioners. To make these relationships more legible we have included a simple visualisation of the topics explored in each conversation at the beginning of each chapter. These 12 topic maps illustrate the relationships between concepts and ideas present in the book. These visualisations should help you navigate the content of the discussions and read for overlap, enabling you to better explore the commonalities and differences found herein. At the end of each chapter there is a short text that connects the some of the key ideas from that conversation to the rest of the book, equal parts summary and hypertext.

Necessarily, the conversations are framed by the temporal, locative and institutional contexts all three participants inhabited at that time. Keir no longer leads the digital department at the San Francisco Museum of Modern Art, which he did during the development of this project and throughout all the recordings. Many other participants have also changed roles since they took part in the project. As such, each chapter begins with the date of the interview and the roles the two participants held when the conversation was recorded, prefaced by a short introduction that provides framing context for the edited transcript that follows.

In the book's foundational essay, Suse draws upon cultural critic and futurist Ziauddin Sardar's theories of "postnormal times" to frame a set of provocations on the digital future of museums that provide a societal, technological and very human context for the 12 conversational chapters that follow.

We conclude by asking, "where do we go from here?" Taking the book's provocations and conversations as both context and a platform for future exploration, we present a set of possible futures and complex questions that are made more intelligible by ideas presented herein and through the work happening in museums around the world that these ideas represent.

In the months after recording and editing these conversations, it became clear that this project has changed the way we think and work. We have a more nuanced, intersectional and comprehensive view of museum practice, and, more specifically, how technology's ever-increasing rate of change is impacting society, our relationships, who and how we trust, how we create, consume and even think, and, in these pages, how it is impacting the museum sector. Along the way we have let go of many of our own preconceived beliefs and embraced new ideas, strategies and tactics.

---

9 Daryl Karp, in Conversation 9.

We encourage you to explore the book, read the essays, take in individual chapters or binge on the whole thing. Search for and make your own links and synthesis between the ideas. There are many connections that can be made, and while we have done some of the work to draw ideas together, there is more to be done to understand and interpret these discussions. The conversations can be read in any order or studied alone. However, when taken together they offer a rich, nuanced and – at times – contradictory view of the digital future of museums. Thanks for reading, and we look forward to hearing how you'll help shape the future of museums!

Professor Keir Winesmith
Dr Suse Anderson

SOME PROVOCATIONS ON THE DIGITAL FUTURE OF MUSEUMS

Dr Suse Anderson

We live in what Ziauddin Sardar calls "postnormal times," characterised by "uncertainty, rapid change, realignment of power, upheaval and chaotic behaviour" (Sardar 2010). Postnormal times are, for Sardar, defined by complexity, chaos and contradictions – nearly impossible to comprehend, highly connected and replete with tensions, leading to uncertainty and risk and bringing into question conventional or normalised modes of thinking and behaviour. Underpinned by Thomas Kuhn's concept of normal paradigmatic science (Kuhn 1962), Sardar's use of "postnormal" is directly derived from work by Jerome R. Ravetz and Silvio O. Funtowicz, who coined the term "post normal science" when working on the mathematics of risk. They observed that while the "normal" model or paradigm of science expects "regularity, simplicity and certainty in the phenomena and in our interventions" (Funtowicz and Ravetz 2001, 15), increased recognition of complexity and uncertainty in scientific work now suggest that "normal" science can no longer be taken for granted. Complex systems are composed of many components (including humans and processes) that interact non-linearly with one another, and change dynamically in response to the environment, which is itself complex. Complexity can be found in nature, social contexts and artificial contexts (such as technologies and institutions) (Cameron and Mengler 2009), and across all of these. In the modern era, humans in Western civilisations sought to control or gain certainty over complexity by creating maps or models that could localise control, (Grinell 2014) making choices about what variables should be included. Because they are reductive by nature, all models of complex systems are necessarily inadequate. However, within the normal modern paradigm, reducing complexity via control and certainty enabled significant techno-scientific progress such that the limitations of such approaches could be ignored (Sardar 2010; Grinell 2014).

This is no longer the case. In expanding these observations beyond science to include other disciplines and society at large, Sardar argues that postnormal times such as these, when the world has been transformed by globalisation and urbanisation, economic crises, demographic changes including generational shifts and substantial technological development, demand that *a priori* assumptions that progress is assured, modernisation is desirable and efficiency is necessary are interrogated. Climate change threatens stability on multiple fronts and the geopolitical landscape is changing simultaneously and quickly in many ways. While political participation is increasing globally, trust in democracy and its associated institutions has been deteriorating. ("Democracy Index 2018: Me Too? Political Participation, Protest and Democracy" 2019; "2017 Edelman Trust Barometer-Executive Summary" 2017) Such changes have been exacerbated by the development of the internet 50 years ago, because the scope, nature and functioning of networks has left the world more connected and interconnected. The network now powers cities, governments, financial markets and public discourse. Although networks are designed to be robust and adaptive, enabling many pathways to route around damage, strikes against the system can have knock-on and multiplying effects, so chaotic behaviour has potentially global impacts. Concurrently, there have been significant shifts to population-level content consumption patterns as news and popular culture is pushed to smartphones and ever-present devices. The dynamics of collective attention are accelerating, fragmenting and hastening public discussion (Lorenz-Spreen et al. 2019). As individuals, institutions and society, we are ill-equipped to deal with, and make sense of, such complexity.

Such transformations are challenging for epistemological institutions, including museums and galleries, whose normal approaches to knowledge were informed by the assumptions of the modern era. Over hundreds of years, starting in the 18th century, knowledge in the museum was created through systematic organisation of the world through objects segmented into defined disciplines. "Normal" practice within museums worked "to legitimate particular views of the natural and cultural world" (Cameron and Mengler 2009, 204), through control; simplifying dense, difficult histories to fit into display cases and exhibition halls, at times representing whole histories, lives or moments in time through a single object, explained by captions with limited word counts. As Nicolas Poole writes:

> For centuries, the role of museums has been to digest complexity and express it as pattern. Whether it is a linear hang in an art museum, giving the impression of a coherent progression of art-historical movements or a social-historical display giving the impression of a singular "community" with identifiably-shared beliefs and values – we are temples to the illusion of order and predictability in a complex and chaotic world.
> *(Poole 2014)*

Yet museums, too, are now situated in postnormal times (Grinell 2014), and this kind of normal practice, which stages extremely limited versions of history within the museum's walls (Rekdal 2014) has increasingly come into question.

Instead, acknowledging complexity becomes essential (Cameron and Mengler 2009, 2010; Gurian 2017; Bennett 2004).

In many ways, museums have been publicly challenged to deal with complexity within their collecting and display practices since at least the 1960s, when postmodernism, structuralism and political movements including the first wave of feminism challenged institutions of knowledge to acknowledge and include pluralistic narratives within their walls. Such movements drew attention to the hegemonic nature of history as presented in museums, which is built upon structures of colonial and patriarchal oppression. While modern grand narratives functioned to strengthen and legitimise existing power relations, these discursive and intellectual frameworks emphasised openness, transparency and polyvocality as strategies to unshackle museums and the histories contained within them from existing systems of power and oppression – arguably to little practical effect. In many parts of the world, museum visitors, staff and volunteers still fail to reflect the diversity of the communities that surround them.[1] In the West, collections and displays continue to reinforce dominant narratives that centre white, male creators and values (Reilly 2018), despite clear evidence that such histories are inadequate at best. These challenges have precipitated meaningful existential angst within the museum sector, as questions arise about the mission, purpose and practicalities of these institutions that are "struggling to survive with a nineteenth-century model in a twenty-first-century world" (Black 2016, 392).

The contradictions between the growing awareness within the sector of the institutional failures of museums to grapple fully with complex and contradictory histories as expressed through their objects is typical of the kinds of contradictions that are central to postnormal times. As Sardar identifies, there is a contradiction between the focus on seemingly unprecedented change in the current era, such as the rapid change and development associated with technology, and the quasi-stasis of in-built systems of oppression and inequality, which become more embedded over time. For instance, early conceptions of cyberspace imagined it as "a world that all may enter without privilege or prejudice accorded by race, economic power, military force, or station of birth" (Barlow 1996). When social media first came to prominence in the mid-2000s, it was lauded as a triumph for democracy. New platforms and accessible tools for the creation of content lowered the barriers to participation in online spaces and brought a massive new influx of participants in public or semi-public online spaces. Not only did this reconfigure notions of publicness (Baym and boyd 2012), as more people began to

---

1 In a 2015 survey of art museum professionals, the Mellon Foundation discovered that only 28% of art museum staff at American Association of Art Museum Director member museums were from minority backgrounds, with most in security, facilities, finance and human resources. Only 4% of curators, educators, conservators and directors are African American and only 3% were Hispanic. By 2018, Mellon's second demographic survey found that art museum staff have become more racially and ethnically diverse, although museum leadership departments showed a comparative lack of diversification (Schonfeld et al. 2015; Westermann et al. 2019).

participate in online discourse, existing models of authorship and authority came into question. Individual creators could bypass traditional gatekeepers of knowledge and access, such as news organisations, governments, universities and museums, and gain visibility and voice that had not previously been possible. During the Arab Spring, for instance, social media offered meaningful tools for political mobilisation and communication (Frangonikolopoulos and Chapsos 2012). Networked technologies have offered individuals and groups ways to communicate and organise, becoming a powerful force for change in both developed and developing countries. However, even as the internet has enabled increased democracy through participatory practices, the major platforms have, "systematically divided people into market segments and political tribes" (Vira 2019). Digital redlining[2] prevents certain users from seeing employment and housing opportunities, while digital inequality can exacerbate educational and income inequality (Rideout and Katz 2016). For all the potential of the internet to create open, transparent and democratic systems of communication, we now witness, "a universal network splintered into a mess of disjointed platforms and disenfranchised user bases by a handful of corporations seeking profit and power" (Vira 2019). Tech giants such as Facebook, Google, uber and Amazon have become so powerful in their specific parts of the market that they are functionally sovereign, "structuring the markets for others" ("Law Professor Frank Pasquale Q&A · Nesta" 2019). Such platforms simultaneously control the platforms and the mechanisms of access, with relatively little regulation or transparency.

Within museums, too, it has long been imagined that networked technologies would offer new kinds of access, effectively democratising the museum and its collections and making them open to all regardless of class, location, education or economic power. However, just as there are thresholds for visitors to the museum to overcome upon entry into its spaces, there are explicit and implicit barriers to entry to online collections, including barriers related to orientation and navigation of online collections, their use of language, their categorisation of objects and limited rich information. There are also contradictions inherent in the liberatory rhetoric often associated with digitisation, and the entrenched biases found in the collection. It is often the case that online collections further embed institutional biases. In many institutions, the underlying corpus of data powering the online collection is fundamentally imbalanced, often skewing heavily white and male. A recent study of online data of more than 40,000 works by over 10,000 artists from 18 major U.S. museums – including the *Metropolitan Museum of Art*, the *National Gallery of Art*, the *Philadelphia Museum of Art*, the *Art Institute of Chicago*, the *Los Angeles County Museum of Art*, and the *Museum of Fine Arts, Houston* – found that 85% of artists in these collections online are white, and 87% are male (Topaz et al. n.d.). It is not merely the artists that are white,

---

2 In the U.S., redlining included both informal and institutionalised policies to prevent certain racial groups from accessing financial resources and mortgages. The effects of redlining continue to be felt today.

either. In January 2018, the Google Arts and Culture Institute released an update that used facial recognition software to compare users to portraits from museums around the world. However, the results for non-white faces were hugely limited, and often included and reinforced harmful stereotypes (Goggin 2018). Journalist Benjamin Goggin attributed the results to the dual factors of Google's selection of its partner institutions, which skewed European and North American, and to the problematic histories of representation within those institutions.

Unfortunately, the experience of the Google Arts and Culture Institute app is not an isolated case. Algorithmic decision-making is sometimes assumed to be more neutral and less open to bias than human decisions. However, algorithms rely on the data they can access and the assumptions built into their code. When that data is flawed, inadequate or biased – as is often the case with museum collections data – or the encoded decisions are intentionally or unintentionally discriminatory, so too are the results of its analysis. As technology companies algorithmically analyse big, connected data, often informed by generations of systematic biases, to supply users with personalised experiences of the world, algorithmic discrimination is creating new forms of racial and class domination, with little to no transparency available to those whose lives are being affected. Recent studies demonstrate that machine learning algorithms can discriminate based on classes like race and gender (Buolamwini and Gebru 2018; Ali et al. 2019). In March 2019, Facebook was sued by the U.S. Department of Housing and Urban Development over the way it let advertisers target ads by race, gender and religion – all protected classes under U.S. law (HUD Public Affairs 2019). The company announced that it would stop allowing this in key categories where federal law prohibits discrimination in ads (Scheiber and Isaac 2019). However, even with neutral targeting parameters, advertising and content delivery can still skew significantly along gender and racial lines (Ali et al. 2019), further entrenching the status quo.

Acknowledging this, some companies are trying to better account for diversity in their datasets in order to better serve non-white publics. For instance, in January 2019, IBM announced the release of a "new large and diverse dataset called Diversity in Faces (DiF) to advance the study of fairness and accuracy in facial recognition technology" (Merler et al. 2019, 2). The dataset provides 1 million human facial images annotated using 10 coding schemes, with the hope to "accelerate the study of diversity and coverage of data for AI facial recognition systems to ensure more fair and accurate AI systems." The photoset was drawn from creative commons licensed images on Flickr. However, as Meredith Whittaker, co-director of the AI Now Institute notes in an interview on NBC (U.S.), "People gave their consent to sharing their photos in a different internet ecosystem" (Solon 2019), and were not necessarily expecting to have their photos used to train facial recognition systems that may eventually be used to surveil them.

Museums, too, have started to engage with surveillance technologies, including monitoring of online behaviours and interactions and using location-awareness technologies in the galleries, to better understand and respond to their audiences. Like other contradictions found in postnormal times, this offers institutions huge benefits as they seek to better serve their audiences, but also offers risks in rendering certain audiences less visible and less discernible. As more aspects of audience behaviour are made visible through such technologies, institutions gain new opportunities to meet their needs and expectations. This can be beneficial to institutions seeking to create better and more personalised experiences for their audiences. As Rob Stein comments:

> I've always felt that technologies and media themselves are not inherently good or inherently evil, but in fact they're just tools that people will employ for good outcomes or not so good outcomes. . . . I don't think there's any reason why museums and culture organisations in general should be afraid of using those tools to do good in the world, to create inclusive dialogues, to highlight the diversity of culture and history, to promote tolerance of different ideas, and to promote acceptance of difference.

Such aims, that seek to leverage the affordances of digital technologies to create better experiences for visitors, and to promote civic discourse and democratic ideals, are laudatory and align with the broad aims of the sector. In doing so, however, it becomes essential for institutions to engage critically with the limitations and ethical questions that such technologies engender. As mentioned above, the algorithms used in conjunction with such technologies can be discriminatory. In mediated contexts, it can be tempting to conflate visible audiences with all audiences. However, "as identity practice becomes explicit, measurable and analysable . . . those who cannot or choose not to participate are increasingly powerless to shape their own experiences or to influence the organisations that serve them" (Anderson 2019). Natalie Kane, Curator of Digital Design at the *V&A Museum*, recently Tweeted her concerns:

> There's a very worrying trend in museums engaging with surveillance technologies and analytics (such as computer vision to count visitor numbers) in the name of "better understanding audiences," or data science from social media to analyse public culture events to sell on . . . to corporate entities to make "art" (read: technology demos) for spaces such as dystopian culture-hole Coal Drops Yard. It needs to stop, immediately. This not only further disadvantages disengaged communities, but essentially turns cultural institutions into R&D for capitalism . . . it's not a fun experiment for developers or "innovation/culture/creative labs" it is an infringement on our duty as civic entities when we monitor our visitors rather than engage with their curiosity, and collect more data than we ever need to, against their benefit.
>
> *(Kane 2019)*

Her concerns echo those of Aaron Cope, who recently argued that:

> The impulse to market, sell or otherwise leverage the data that we as museums collect through location-based technologies will be strong even though this has nothing, and I mean literally nothing, to do with the business of cultural heritage. The impulse to develop museum exhibitions solely for the purpose of collecting visitor data will surely follow, which assumes someone isn't already doing it.
>
> *(Cope 2018)*

Museums have increasingly come to rely on visitor data to understand the success of their enterprise and to demonstrate to funders and other stakeholders the impact of their work. But they are not the only institutions moving towards a data-driven business model. As big, connected data has become fundamental to the networked information infrastructure, Shoshana Zuboff has argued that it is becoming the foundational component in a new logic of accumulation, dubbed "surveillance capitalism" (Zuboff 2015). Surveillance capitalism is a form of information capitalism that "aims to predict and modify human behaviour as a means to produce revenue and market control" (Zuboff 2015, 75). She writes:

> With the new logic of accumulation that is surveillance capitalism, a fourth fictional commodity emerges as a dominant characteristic of market dynamics in the 21st century [alongside labour, real estate and money]. Reality itself is undergoing the same kind of fictional metamorphosis as did persons, nature, and exchange. Now "reality" is subjugated to commodification and monetisation and reborn as "behaviour." Data about the behaviours of bodies, minds, and things take their place in a universal real-time dynamic index of smart objects within an infinite global domain of wired things. This new phenomenon produces the possibility of modifying the behaviours of persons and things for profit and control. In the logic of surveillance capitalism there are no individuals, only the world-spanning organism and all the tiniest elements within it.
>
> *(Zuboff 2015, 85)*

Whereas the modern era was defined by its attempt to control and reduce variables in order to better understand the world, the era of Big Data instead collects and connects reams of data – big and small – in order to algorithmically mine such data for patterns across human behaviour. Online, all actions leave a trace, whether conducting Google searches or swiping right on Tinder. Our objects and spaces are quantified, too. With little regard for user privacy, all such data can be captured, kept, combined and mined for new understanding and insight. Yet, as danah boyd and Kate Crawford caution, automating research in this way, "reframes key questions about the constitution of knowledge, the processes of research, how we should engage with information, and the nature and the categorisation of reality" (boyd and Crawford 2011, 3). This aligns with the second contradiction that Sardar spells out, concerning knowledge in postnormal times, wherein knowing more, as we do from research across different fields, has not led to a decrease in ignorance, but has rather increased it. He writes, "While we are bombarded with information on almost all and every subject, we have very limited capability to actually discern what is important and what is trivial" (Sardar 2010).

Even as social media enabled significant growth in user-generated content, the mechanisms for the verification of information online were not linked to its truth or factuality. Rather, platforms such as Google, Facebook and YouTube created recommendation engines that used complex algorithms to define the value of a resource for its users, with little appetite for policing content for veracity. The long tail of that decision has created an online context where bots mimic humans and troll farms exploit the features of social media platforms to deliberately spread "sensationalist, conspiratorialist, and other forms of junk political news and misinformation to voters across the political spectrum" ("Russia Used Social Media for Widespread Meddling in U.S. Politics: Reports" 2018). At the same time, machine learning is helping create a world where real and fake are increasingly indistinguishable to the human eye. For instance, deepfakes – a kind of algorithmically doctored video – use generative adversarial networks to alter the appearance of source images or video to create new videos and images that look and sound real. It is becoming harder to believe our own eyes or to know who to trust.

Ironically, as information has proliferated and trust in traditional knowledge institutions and gatekeepers has eroded, many individuals now "do their own research" rather than seeking the advice of experts (Dimock 2019). Such research requires the individual to undertake significant cognitive responsibility over their own decision-making, often in areas that are not related to their expertise, while personalisation practices shapes and limit the information the individual is exposed to. danah boyd argues that media literacy has backfired in the digital age, in part because critical thinking and critical questioning have been weaponised and because each person is now expected to decide whether each new piece of information we encounter is true or not (boyd 2017, 2018). This is a condition that Seph Rodney draws attention to, in the first discussion of this book, suggesting that:

> on a social level, that the tendency to imagine that there is no authority out there which one must count on, has left us rudderless. There's a direct correlation with the internet, and the notion that the truth is out there and that you, wherever you are, you can be empowered to find it and that people are hiding it from you, but you can discover it yourself. We think that agency provides us with knowledge, and it doesn't.

Without robust institutionalised mechanisms for verifying facts and truth online, misinformation spreads easily. But the challenges are more pernicious than that. Writer Britney Gil argues we are in a crisis of epistemology, in which "the ways we come to know the world [are] bound up in collapse of trust in fundamental institutions" (Gil 2016). Cory Doctorow makes this point even more forcefully:

> we're not living through a crisis about what is true, we're living through a crisis about how we know whether something is true. We're not disagreeing about facts, we're disagreeing about *epistemology*. The "establishment" version of epistemology is, "We use evidence to arrive at the truth, vetted by independent verification (but trust us when we tell you that it's all been independently verified by people who were properly skeptical and not the bosom buddies of the people they were supposed to be fact-checking)."
>
> The "alternative facts" epistemological method goes like this: "The 'independent' experts who were supposed to be verifying the 'evidence-based' truth were actually in bed with the people they were supposed to be fact-checking. In the end, it's all a matter of faith, then: you either have faith that 'their' experts are being truthful, or you have faith that we are. Ask your gut, what version *feels* more truthful?"
>
> (Doctorow 2017)

For museums, as institutions of knowledge and democracy, this crisis in epistemology is – and must be – deeply unsettling. Many of the once-trusted institutional mechanisms for establishing truth and veracity at a societal level have come into question in light of deep complexity, even as sanctioned and institutional narratives have been contested. Although museums have retained significant levels of public trust compared to many other institutions, institutional systems are linked laterally and vertically, so collapsing trust impacts all, if not in the same way or at the same time. Additionally, institutional trust has a broad impact on social trust, or trust in generalised, abstracted others, which promotes desirable collective outcomes (Sønderskov and Dinesen 2016), so museums have good reason to be concerned about falling levels of trust.

It is unsurprising, therefore, that concerns about trust surface often within this book. Adriel Luis wonders about the purpose of trust in museums. He asks, "How do we cultivate museums as environments where people can have the kinds of conversations that they might not have anywhere else?"[3] Daryl Karp, too, speaks about the challenges of creating "civil, engaged dialogue,"[4] when such challenging conversations often require people to have a pre-existing relationship. Meanwhile, Seph Rodney suggests that,

---

3 Adriel Luis, in Conversation 4
4 Daryl Karp, in Conversation 9

> if we don't learn to manage controversy in a way that is not panicked, that doesn't imagine that conversations should always be pleasant and nurturing and comforting, then museums are going to ultimately fail in the task of being the kind of institutions that can be trusted.[5]

Frequently, conversations about trust are linked to questions about museum neutrality, and the kinds of curatorial choices that museums make in their exhibition, programming and collecting practices – particularly when communities have not previously had positive experiences with the museum. Luis continues:

> Museums are historically instruments of colonisation, particularly in non-Western regions. The first museums in Asia were in India, founded by the British, and in Indonesia, founded by the Dutch. These museums were founded as tools to present narratives to people of colour via their colonisers. How can we move forward, assuming that nowadays museums present themselves as public spaces . . . and under the mis-assumption that everybody trusts museums? These issues that people have with museums haven't been solved.[6]

For generations, museums have benefitted from a mythology, borne of a Eurocentric Western knowledge system, "whose most basis [sic] tenet was the delusional notion that they were value neutral, universal and inherently good" (Sardar 2010). Against this background, First Nations knowledge production has been delegitimised and historical and cultural pluralism has been denied (Sentance 2018). While notions of museum neutrality have been interrogated for decades, this mythology has come under renewed public scrutiny in the wake of social media, as countless new people, including museum professionals who had previously been voiceless, have been able to enter the public discourse and share their stories and experiences, which often contradict canonical or official narratives.[7]

Digital technologies have, in some ways, prompted museums to address the contradictions inherent in their work, and the challenges of complexity. For instance, although the exhibition is the most visible example of the mechanisms that museums use to defang complexity through reduction, it is hardly the only one. Museum collection documentation systems also promote regularity, simplicity and certainty through categorisation (Cameron and Mengler 2009). As a tool for institutional persistence over time, such systems limit flexibility and difference by flattening all objects into predetermined comparative frameworks

---

5 Seph Rodney, in Conversation 1
6 Adriel Luis, in Conversation 4
7 See, in particular, the #MuseumsAreNotNeutral movement, founded by LaTanya S. Autry and Mike Murawski (Autry 2017; Murawski 2017).

of classification. Classification defines relationships between objects, and frames how they will be read and understood within the context of the institution. However, as LaToya Devezin and Barbara Makuati-Afitu make clear in Conversation 2 in this book, such pre-existing systems built around Western frameworks of knowledge are challenged by the need to deal with Indigenous and non-Western systems of knowledge. The foundational assumptions upon which museum collections systems have been founded restrict and limit the kinds of knowledge that are stored within them. This kind of cultural encoding places all objects within a kind of normative framework that legitimises and prioritises some forms of knowledge over others. Such systems are also challenged through the act of putting collections online, and reconnecting museum objects and knowledge to global flows of culture and information. The networked object can be connected to, and co-opted into, countless narratives, including narratives and purposes that are competing and contradictory, many of which will not be defined or determined by the museum. These actions bring into question museum narratives and choices, and emphasise the plurality of possible uses and outcomes for museum objects far beyond the institution. As I have previously written, "The networking of knowledge is inviting the museum to rethink its models of expertise, and to further consider the role of other contributors in the creation of museum knowledge" (Cairns 2013, 115).

Although as yet under-explored and under-exploited, leaning into and embracing complexity also offers important opportunities to better align the museum's internal practices and approaches to knowledge work with the rhetoric surrounding participation and polyvocality that has proliferated within the sector for decades. Indeed, Klas Grinell proposes that postnormal times offer an opportunity for museums. While Grinell acknowledges that the elimination of a "common modern social imaginary" might be disconcerting, he suggests that, "the future must be pluriversal; it must be built from the understanding that no single paradigm can understand and order the world" (Grinell 2014, 173). He proposes that public museums should seek to create space for different and even clashing experiences of phenomena, in order to be a space for deliberative democracy. Rather than seeking to minimise different, therefore, complexity, chaos and contradictions become something to be acknowledged and accepted, where our differences and commonalities intersect with one another, and are localised and situated. Or, as Sardar writes:

> To accept that there are no right and wrong answers does not mean we abandon the search for truth or solutions but it does entirely change the process and kind of objectives we set for our endeavours. When there are no right or wrong answers everyone, every perspective, has a contribution to make, anyone is as likely as another to have some part of a potential solution.
>
> *(Sardar 2010)*

As institutions, museums have long, troubling histories as mechanisms of imperialism and colonisation. However, these histories are not past. Although new technologies are sometimes seen as an answer to old problems, what this book makes clear is that technology problems, too, are human problems, where power and relationships are built upon historical biases and inequities already inherent in our systems and data. Yet the context within which museums now work is one assembled upon and of networks, and they are therefore at the mercy of network effects. When one museum puts its collections online, there may indeed be a democratising effect, as objects previously hidden become visible and usable, and the networked object is inserted into the cultural flows of the network. When hundreds or thousands of museums digitise their collections and put them online, the effect is one that further embeds the status quo at scale, making claims writ large across millions of objects about whose histories and objects have been worth collecting, and in what circumstances, as part of whose narratives.

The conversations in this book speak to multiple strategies that museums are employing to be inclusive of diverse perspectives and responsive to their communities. Digital technologies have offered museums a new set of tools and mechanisms through which to start thinking about and dealing with postnormal times. However, they have also brought a new context to the work, one replete with complexity, contradictions and chaos. Such context asks digital practitioners within museums to move beyond concerns about innovation and institutional progress in isolation, and prompt them to think about the long- and short-term impact of the work they do, and how it contributes to or challenges existing power relations. While the challenges that postnormal times bring do not only affect the museum's digital work and output, it is at the many touchpoints where the museum meets the network that these issues are often made most visible.

Complexity and contradictions are not going to disappear from our work as we move in the digital future of museums. Digitising collections and putting them online offers museums new ways to be relevant to the world, even as they reinforce dominant histories and discourses. Surveillance technologies that enable better understanding of audiences and their needs can benefit museums and museum audiences, but they can also carry risks, such as when that data is not collected or cared for with an eye to safety and security, or when algorithms discriminate according to race or gender. We cannot run from these challenges. Instead, we need to find new ways to work that take such contradictions into account and make decisions that, as Laura Raicovich says, "operationalise our values" (Battaglia 2018), and align our institutional body language with our rhetoric about the kinds of institutions we want museums to be. As Seb Chan wrote in notes from a recent speech:

> We need museums to play a new role. And a role that will be shaped largely through intentional design and strategic decision making. We know that

museums are not neutral and that they need to take specific intentional choices to make change.

(Chan 2019)

Perhaps LaToya Devezin put it best when she said, "All stories deserve to be told." It is now incumbent upon those of us concerned with digital practice in museums to figure out how to best do so.

References

"2017 Edelman Trust Barometer-Executive Summary." 2017. https://www.edelman.de/fileadmin/user_upload/Studien/2017_Edelman_Trust_Barometer_Executive_Summary.pdf.

Ali, Muhammad, Piotr Sapiezynski, Miranda Bogen, Aleksandra Korolova, Alan Mislove, and Aaron Rieke. 2019. "Discrimination through Optimization: How Facebook's Ad Delivery Can Lead to Skewed Outcomes," April. https://web.archive.org/web/20190712191030/**https://arxiv.org/abs/1904.02095**.

Anderson, Susan. 2019. "Visitor and Audience Research in Museums." In *The Routledge Handbook of Museums, Media and Communication*, edited by Kirsten Drotner, Vince Dziekan, Ross Parry, and Kim Christian Schrøder, 1st ed., 80–95. Abingdon, OX: Routledge.

Autry, La Tanya. 2017. "Changing the Things I Cannot Accept: Museums Are Not Neutral." Artstuffmatters. 2017. https://web.archive.org/web/20190712191207/**https://artstuffmatters.wordpress.com/2017/10/15/changing-the-things-i-cannot-accept-museums-are-not-neutral/**.

Barlow, John Perry. 1996. "A Declaration of the Independence of Cyberspace." Electronic Frontier Foundation. 1996. https://web.archive.org/web/20190712191309/**https://www.eff.org/cyberspace-independence**.

Battaglia, Andy. 2018. "The ARTnews Accord: Aruna D'Souza and Laura Raicovich in Conversation." Art News. 2018. https://web.archive.org/web/20190712191409/**http://www.artnews.com/2018/05/14/artnews-accord-aruna-dsouza-laura-raicovich-conversation/**.

Baym, Nancy K., and danah boyd. 2012. "Socially Mediated Publicness: An Introduction." *Journal of Broadcasting & Electronic Media*. Abingdon, OX: Taylor and Francis Group, LLC. http://www.tandfonline.com/doi/full/10.1080/08838151.2012.705200

Bennett, Tony. 2004. "The Exhibitionary Complex." In *Grasping the World: The Idea of the Museum*, edited by Donald Preziosi and Claire Farago, 413–42. Hants, UK & Burlington, VT: Ashgate.

Black, Graham. 2016. "Remember the 70%: Sustaining 'Core' Museum Audiences." *Museum Management and Curatorship* 31 (4): 386–401. https://doi.org/10.1080/09647775.2016.1165625.

boyd, danah. 2017. "Did Media Literacy Backfire?" Data & Society: Points. 2017. https://web.archive.org/web/20190712191604/**https://points.datasociety.net/did-media-literacy-backfire-7418c084d88d?gi=3f5228e95dae**.

———. 2018. "You Think You Want Media Literacy… Do You?" Data & Society: Points. 2018. https://web.archive.org/web/20190712191646/**https://points.datasociety.net/you-think-you-want-media-literacy-do-you-7cad6af18ec2?gi=8b52512e5eaa**.

boyd, danah, and Kate Crawford. 2011. "Six Provocations for Big Data." *Oxford Internet Institute's "A Decade in Internet Time: Symposium on the Dynamics of the Internet and Society.*" Oxford, UK. https://web.archive.org/web/20190713073737/**https://papers.ssrn.com/sol3/papers.cfm?abstract_id=1926431**.

Buolamwini, Joy, and Timnit Gebru. 2018. "Gender Shades: Intersectional Accuracy Disparities in Commercial Gender Classification *." In *Proceedings of Machine Learning Research*, edited by Sorelle A. Friedler and Christo Wilson, 81:1–15. Conference on Fairness, Accountability, and Transparency. http://proceedings.mlr.press/v81/buolamwini18a/buolamwini18a.pdf.

Cairns, Susan. 2013. "Mutualizing Museum Knowledge: Folksonomies and the Changing Shape of Expertise." *Curator: The Museum Journal* 56 (1): 107–19.

Cameron, Fiona, and Sarah Mengler. 2009. "Complexity, Transdisciplinarity and Museum Collections Documentation: Emergent Metaphors for a Complex World." *Journal of Material Culture* 14: 189–218.

———. 2010. "Activating the Networked Object for a Complex World." In *Handbook of Research on Technologies and Cultural Heritage: Applications and Environments*, edited by Georgios Styliaras, Dimitrios Koukopoulos, and Fotis Lazarinis, 166–87. Hershey, PA: IGI Global. https://doi.org/10.4018/978-1-60960-044-0.

Cope, Aaron Straup. 2018. "Successful Distractions." [This Is Aaronland]. 2018. https://web.archive.org/web/20190712191945/**https://www.aaronland.info/weblog/2018/09/12/distractions/**.

"Democracy Index 2018: Me Too? Political Participation, Protest and Democracy." 2019. London. https://web.archive.org/web/20190630082129/**https://www.eiu.com/public/topical_report.aspx?campaignid=Democracy2018**.

Dimock, Michael. 2019. "An Update on Our Research into Trust, Facts and Democracy." Pew Research Center. 2019. https://web.archive.org/web/20190712192126/**https://www.pewresearch.org/2019/06/05/an-update-on-our-research-into-trust-facts-and-democracy/**.

Doctorow, Cory. 2017. "Three Kinds of Propaganda, and What to Do about Them / Boing Boing." BoingBoing. 2017. https://web.archive.org/web/20190712192224/**https://boingboing.net/2017/02/25/counternarratives-not-fact-che.html**.

Frangonikolopoulos, Christos A., and Ioannis Chapsos. 2012. "Explaining the Role and the Impact of the Social Media in the Arab Spring." *Global Media Journal: Mediterranean Edition* Fall: 10–20. https://web.archive.org/web/20171130171857/**http://www.academia.edu/2370755/Explaining_the_role_and_impact_of_social_media_in_the_Arab_Spring_**.

Funtowicz, Silvio, and Jerry Ravetz. 2001. "Post-Normal Science. Science and Governance under Conditions of Complexity." In *Interdisciplinarity in Technology Assessment*, 15–24. Berlin, Heidelberg: Springer Berlin Heidelberg.

Gil, Britney. 2016. "Texts, Facts, Emotions, and (Un)Making a Nation – Cyborgology." Cyborgology. 2016. https://web.archive.org/web/20190712192418/**https://thesocietypages.org/cyborgology/2016/12/02/texts-facts-emotions-and-unmaking-a-nation/**.

Goggin, Benjamin. 2018. "Is Google's Arts And Culture App Racist? – Digg." Digg. 2018. http://web.archive.org/web/20190712192454/**http://digg.com/2018/google-arts-culture-racist-face**.

Grinell, Klas. 2014. "Challenging Normality: Museums In/As Public Space." In *Museums and Truth*, edited by Annette B. Fromm, Per B. Rekdal, and Viv. Golding, 169–188. Newcastle upon Tyne, England: Cambridge Scholars Publishing.

HUD Public Affairs. 2019. "HUD Charges Facebook with Housing Discrimination Over Company's Targeting Advertising Practices." Washington, DC: US Department of Housing and Urban Development. https://web.archive.org/web/20190712192623/**https://www.hud.gov/press/press_releases_media_advisories/HUD_No_19_035**.

Kane, Natalie D. 2019. "No Title." Twitter. https://web.archive.org/web/20190712192728/**https://twitter.com/nd_kane/status/1144195380076302336**.

"Law Professor Frank Pasquale Q&A · Nesta." 2019. Finding Ctrl. 2019. https://web.archive.org/web20190712193057/**http://findingctrl.nesta.org.uk/frank-pasquale/**.

Lorenz-Spreen, Philipp, Bjarke Mørch Mønsted, Philipp Hövel, and Sune Lehmann. 2019. "Accelerating Dynamics of Collective Attention." *Nature Communications* 10 (1): 1759. https://web.archive.org/web/20190614124801/**https://www.nature.com/articles/s41467-019-09311-w**.

Merler, Michele, Nalini Ratha, Rogerio S. Feris, and John R. Smith. 2019. "Diversity in Faces," January. https://web.archive.org/web/20190712193151/**https://arxiv.org/abs/1901.10436**.

Murawski, Mike. 2017. "Museums Are Not Neutral." Art Museum Teaching. 2017. https://web.archive.org/web/20190712193214/**https://artmuseumteaching.com/2017/08/31/museums-are-not-neutral/**.

Poole, Nicholas. 2014. "Change." CODE | WORDS: Technology and Theory in the Museum – Medium. 2014. https://web.archive.org/web/20190712193259/**https://medium.com/code-words-technology-and-theory-in-the-museum/change-cc3b714ba2a4**.

Reilly, Maura. 2018. *Curatorial Activism: Towards an Ethics of Curating*, 1st ed. London: Thames & Hudson.

Rekdal, Per B. 2014. "Why a Book on Museums and Truth?" In *Museums and Truth*, edited by Annette B. Fromm, Per B. Rekdal, and Viv. Golding, 1st ed., ix–xxv. Newcastle upon Tyne, England: Cambridge Scholars Publishing.

Rideout, Victoria, and Vikki S. Katz. 2016. "Opportunity for All? Technology and Learning in Lower-Income Families." www.vikkikatz.com.

"Russia Used Social Media for Widespread Meddling in U.S. Politics: Reports." 2018. Reuters. 2018. https://web.archive.org/web/20190712193530/**https://www.reuters.com/article/us-usa-trump-russia-socialmedia/russia-used-social-media-for-widespread-meddling-in-u-s-politics-reports-idUSKBN1OG357**.

Sardar, Ziauddin. 2010. "Welcome to Postnormal Times." *Futures* 42 (5): 435–44. https://web.archive.org/web/20190712193649/**https://ziauddinsardar.com/articles/welcome-postnormal-times**.

Scheiber, Noam, and Mike Isaac. 2019. "Facebook Halts Ad Targeting Cited in Bias Complaints." NYTimes.Com, March 19, 2019.

Schonfeld, Roger, Mariët Westermann, and Liam Sweeney. 2015. "The Andrew W. Mellon Foundation Art Museum Staff Demographic Survey." https://web.archive.org/web/20190712193740/**https://mellon.org/programs/arts-and-cultural-heritage/art-history-conservation-museums/demographic-survey/**.

Sentance, Nathan "Mudyi." 2018. "Your Neutral Is Not Our Neutral." Archival Decolonist [-O-]. 2018. https://web.archive.org/web/20190712193826/**https://archivaldecolonist.com/2018/01/18/your-neutral-is-not-our-neutral/**.

Solon, Olivia. 2019. "Facial Recognition's 'Dirty Little Secret': Millions of Online Photos Scraped without Consent." NBC News. 2019. https://web.archive.org/web/201907 12194056/**https://www.nbcnews.com/tech/internet/facial-recognition-s-dirty-little-secret-millions-online-photos-scraped-n981921**.

Sønderskov, Kim Mannemar, and Peter Thisted Dinesen. 2016. "Trusting the State, Trusting Each Other? The Effect of Institutional Trust on Social Trust." *Political Behavior* 38 (1): 179–202. https://web.archive.org/web/20161203201250/**http://link.springer.com/article/10.1007/s11109-015-9322-8**.

Topaz, Chad M., Bernhard Klingenberg, Daniel Turek, Brianna Heggeseth, Pamela E. Harris, Julie C. Blackwood, C. Ondine Chavoya, Steven Nelson, and Kevin M. Murphy. n.d. "Diversity of Artists in Major U.S. Museums." Accessed June 11, 2019. https://arxiv.org/pdf/1812.03899.pdf.

Vira, Udit. 2019. "A Field Guide to the Living Internet · Nesta." Finding Ctrl. 2019. https://web.archive.org/web/20190606035455/**https://findingctrl.nesta.org.uk/field-guide-to-the-living-internet/**.

Westermann, Mariët, Roger Schonfeld, and Liam Sweeney. 2019. "Art Museum Staff Demographic Survey 2018." https://web.archive.org/web/20190711212646/**https://sr.ithaka.org/publications/interrogating-institutional-practices-in-equity-diversity-and-inclusion/**.

Zuboff, Shoshana. 2015. "Big Other: Surveillance Capitalism and the Prospects of an Information Civilization." *Journal of Information Technology* 30 (1): 75–89. https://web.archive.org/web/20190331030357/**https://journals.sagepub.com/doi/10.1057/jit.2015.5**.

Conversations on the digital future of museums

CONVERSATION 1

Seph Rodney + Robert J. Stein

". . . authority is always derived from the outside-in, not the other way around."
— *Rob Stein*

FIGURE 1.1

We initially encountered Robert J. Stein (U.S.) through his work as the director of the IMA Lab at the *Indianapolis Museum of Art*, one of the most successful museum digital teams in the field at the time. Years later, and working for a different museum, he gave a fascinating, data-rich conference presentation about the *Dallas Museum of Art*'s data-driven, gamified, free-membership program DMA Friends, which influenced how we both think about the active use of personalisation and data in the museum context. Robert's work consistently interrogates how people use technology in public spaces, especially in museums.

Seph Rodney (Jamaica/U.S.) researched the *Tate Modern* in his Ph.D. thesis and writes about museums as an editor and writer at the influential arts blog *Hyperallergic*. We had independently been reading Seph's writing on *Hyperallergic*, some of which is informed by his Ph.D. research into the customisation and personalisation of the museum visitor experience, and which we felt had a strong relationship to Robert's museum work.

We brought Robert and Seph together to explore the interconnections and points of distinction between Robert's practical experience and museum professional development work at the American Alliance of Museums (AAM) and Seph's research and writing.

This conversation was recorded on January 25, 2018. At the time of the interview, Seph Rodney was an editor for the *Hyperallergic* blog and an adjunct faculty member at Parsons School of Design in New York, U.S. and Robert J. Stein was Executive Vice President and Chief Program Officer of the AAM in Washington, DC, U.S.

KEIR: I'd like to start by looking at how the relationship between audiences and art museums has shifted over the last few decades in response to many socioeconomic, technological and marketing trends, both out in the world and inside museums themselves.

Seph, you've written about how museums, having come into existence in the 18th century, have transitioned through multiple phases, beginning with privileged experiences, to more recent forms of personalised experiences. But I rarely consider myself having had a genuinely personalised experience in museums. Like many visitors, I like to guide myself rather than be guided. However, even when on a guided experience, I feel it's more like "choose your own adventure" than personalisation. Seph, when you think about personalisation, what do you actually mean?

SEPH: There is a specific matrix of relationships which has the visitor at the centre. Looking at museums as an expression of a set of interrelated departments with professional approaches to their work, it seems to me, the curatorial department and the marketing department have, in the last 25 to 30 years, joined forces. Now, when curatorial thinks about staging an exhibition, and they think about what kinds of art, or what audiences they want to have attend, they think in terms of the marketing. What kinds of invitations are they crafting for visitors? I think fundamentally, across the board for museums who are moving towards this paradigm – and not all of them are there yet – is the notion that marketing has to take the lead. You have to find out what visitors want and look to create public programming and exhibitions that will draw larger and more diverse audiences.

What that means is that the position of visitor researcher comes to the forefront of marketing efforts. What museums, especially art museums – I specifically studied art museums in the U.S. and the U.K. – are doing is taking research on the visitor, which actually has a very consumerist slant. They're not just asking questions like what "what did you get from your last visit?" They're asking questions like "what do you need from the next visit?" When I talk about personalised experience, I mean that that information is consistently being mined from visitors, and it's feeding into curatorial and marketing efforts. That's why I think you're getting more programs that are crowd-sourced, more programs that are co-curated. And even though you may not feel personally, Keir, that you have an experience that's personalised, I think that what is happening is that you feel that you have the agency to move through a museum under your own direction, and in some ways that's the fruit of all of these museological changes. Does that make sense?

KEIR: That does, and it ties really closely to what I heard when I listened to Rob's talk at the Re-imagining the Museum conference in Medellín, Columbia. Rob, you've been thinking about these historical phases and transformations in view of the American museum sector. Does Seph's summary match with what you're seeing, or is there more change inside institutions than is reflected outwards?

ROB: In reality, I think it's quite complex. I joined museums in 2005, which is approximately the time when Americans were waking up to personal digital technology, and all of the changes that that afforded in communication and learning and access to media, as well as now the corruption of those cycles. Museums are now wondering what's happening to their relevance, which causes them to ask, in many cases, good questions which have resulted in many museums being more open to what Seph was talking about and really being visitor-centred and researching and studying what visitors want to consume in a museum experience. That's where a lot of the marketing tie-ins come into play.

What I worry about is that the pendulum swings inevitably between three poles. One being an educational pole; museums as a place for the public to learn, which has been true from their founding through today. Another pole being of marketing and popular experience; learning about and creating what contemporary audiences seek and demand, which may be outside the museum's direct areas of interest. A third pole is the scholarly role of museums and the way in which knowledge creation happens. It makes a funny triangle and I think it's relatively easy for museums to jump all-in to any one of those three, while unintentionally neglecting the other. I think it's rather difficult to find a balancing point in the middle. My feeling is that the only healthy place for museums to exist is on the tip of that needle balancing between all three of poles, because when we skew too far to any one direction, I think we see a whole variety of maladies.

SEPH: Rob, what for you constitutes a malady? What's the worst-case scenario? Let's say that the museum doesn't strike a balance among those three choices, what happens for you that's bad?

ROB: When we focus too much on the educational aspects, that the museum is the source of all knowledge and we're going to tell you what to think about the world. If we balance too far on the scholarly aspect, then the interests and motivation of the public doesn't even enter our thinking. And, honestly, if we balance too far on the popular aspect, then you can easily lose what distinguishes a museum from any other entertainment venue. There are places for amusement and entertainment that are purely driven by what can sell, and not at all driven by a point of view that might provoke or push the edges of what the public's comfort zones and interest.

SEPH: Rob, I don't think that that's a controversial point of view at all. I think that a lot of people have been saying that they don't want museums to be entertainment venues, and I think that's absolutely right. I think that museums do play an important role in having us think about our thinking, publicly. I'm with you on that. What I don't feel that you got to in your answer was why the first scenario is bad. Let's say that the museum just wants to be that kind of educational institution that treats the visitor as ignorant pupil. It's been really successful for... it's not an exaggeration to say ... hundreds of years. Why is that the worst-case scenario?

ROB: That's a good question. I think this relates back to the larger social changes of how we consume and interact with media in general. One of the things that learning theory tells us is that the most effective ways to learn are often self-motivated. When we are very curious in and of our own purposes, that can create optimal learning conditions – both for knowledge growth and retention, but also for enjoyment. I think that outside of a formal educational setting, schools or university, there's an expectation that learning can be enjoyable and self-directed. That's a change of educational philosophies in museums over the last two decades, I would say.

SEPH: That's exactly right. That's exactly what my research is on; that set of changes to museums so they become places that welcome self-directed visitors. It's that tilt towards this idea that, in order for people to learn, we need to figure out what the optimal conditions for their learning are, and one of the things that you mentioned, that educational theory has found, is that people do need to realise their own potential in that moment. They need to figure out whether they learn kinaesthetically, or by repetition, or visually, or whatever. I think that that's what's provoked the shift towards a visitor-centred institution.

ROB: I might add that the democratisation of media and voice has led to a general decentralisation of authority structures. I think that includes the authority structure of the museum and institutes of education more generally. What is perhaps behind this expectation of self-directed learning is this decentralisation (and rejections) of "the all-knowing authority." That is no longer as valued as it once was and, in fact, there is more value placed on the intrinsic or beneficial impacts of learning. In my opinion, that's a healthy shift.

SEPH: I think at the AAM's annual conference two years ago you had a keynote speaker talk about how public faith in institutions has just fallen away. Looking at the post-war era, there's this precipitous decline in faith in institutions, faith in government, faith in institutions of higher learning, faith in museums. I think a lot of the fights that have – yeah, I actually do want to call them fights or struggles – have to do precisely with realising that institutions are not neutral and that they are political and that their politics do matter and they do impact people's lives. And that if you want to be heard, recognised and validated publicly, you actually have to fight for such recognition – in and through these institutions. That leads us to, what I think is a cul de sac, because we end up feeling that if institutions like museums are not neutral, and that they are subject to political machinations, then that lack of authority leaves us a little bit rudderless. Would you agree with that?

ROB: In general, I agree with your take on the situation. I think regarding authority, the key thing that has changed is that it's becoming less easy to retain authority, and authority is much more so attributed or given by communities to organisations. There are institutions, organisations, movements, that derive their authority from a grassroots attribution of that authority – as opposed to what may have been true a decade or more ago where institutions claimed and placed a stake in the ground that "we are the authority." That doesn't really work anymore, and I think museums still struggle a little bit with that. But the truth is that authority is always derived from the outside-in, not the other way around. I think you're correct that there's an awakening to the fact that organisations, institutions and museums are not neutral, either by our explicit statements, or our implicit passivity, or non-statement – those are all political movements and choices. That may have always been true, but now our communities and our audiences have a more nuanced recognition and expectation around that.

SEPH: Absolutely.

ROB: I don't know that that makes us rudderless . . .

SEPH: No, no, no. I'm sorry, Rob. I'm not suggesting that museums are rudderless. Rather, I'm suggesting – and this is a sweeping claim – that on a social level, that the tendency to imagine that there is no authority out there which one must count on, has left us rudderless. There's a direct correlation with the internet, and the notion that the truth is out there and that you, wherever you are, can be empowered to find it and that people are hiding it from you, but you can discover it yourself. We think that agency provides us with knowledge, and it doesn't.

KEIR: I think within this framework, it's important to note that this massive reduction in trust in institutions, especially public institutions, is not evenly felt. And especially that libraries and museums as some of the only institutions that still retain trust with over 50% of people, at least in the West. People believe in their expertise, if not necessarily their authority. That brings me back towards where we started. When thinking about risks or pitfalls of real personalisation, it reminds me of internet filter bubbles. The internet's extreme personalisation and extreme filtering.

When I think about organisations reacting to contemporary moments instead of responding to them, I'm reminded of something the artist Hank Willis Thomas said; "museums should not be reactionary, they should be visionary."[1] Is it possible for the personalisation, the customisation of the experience to actually increase polarisation in communities where it's already difficult to create a shared frame of reference for conversation or common political ground? Could studying the audience and moving to one of the tips of your triangle, Rob, lead to institutions that are pre-polarised? This is a pro-gun museum, and this is an anti-gun museum! Or can museums actually use the trust that they still retain and be creators and promoters of spaces for sharing, for discussion and for healing in this moment?

ROB: If I strip all the technology and digital media away and think about the very best experiences I've had in museums, or cultural experiences in general, it's almost always with another person who has knowledge, or a unique point of view, or an interesting perspective where I get into a discussion with them. If you've ever been lucky enough to go through a museum with an artist, or with the director, the curator, you connect more with natural conversation and stories. Your personhood, your thinking and the knowledge that you bring to the museum is integral to the conversation that comes out.

Now thinking about what could museums do. That's not as much about leaning back on their authority, as acknowledging that the museum possesses expertise and knowledge. If we're successful in integrating a visitor's own expertise and knowledge with our own, we can provide a tremendously personal, and I would say enjoyable, learning experience. That would be really valuable and wouldn't succumb to the filter bubble. Figuring out how to do that at scale is where I believe and hope that media and technology can actually help. But, it's incredibly difficult.

KEIR: Seph, you've written a wonderful line about the *Tate Modern*. "It behaved as a romantic object, never fully pleasing the visitor, while not fully frustrating desire either." You're referring to this clever systematic consumerisation

---

1 Hank Willis Thomas, in round table discussion with *SFMOMA* staff (2017).

of the *Tate* visit. From your research at the *Tate*, what can other institutions learn from and build on, given the context Rob provided?

SEPH: I just had a conversation with my colleague at *Hyperallergic* about this. She went to a museum in Texas and said that one of the things that she liked about the museum was that they had the security guards engaging the visitors, asking them questions or trying to answer questions that they elicited from visitors about the work they were seeing. What they've done is they made a hybrid security-docent. That's actually one of the things that I trace back to *Tate Modern*. When *Tate Modern* opened, they took that approach. What museums have to do, art museums particularly, is to nurture conversation in the space of the museum. That could be using guards as docents, in a pre-determined way, to engage with visitors to get the conversation going.

KEIR: The first museum I worked at was the *Museum of Contemporary Art* in Sydney, which is led by Liz-Anne McGregor, and she had the same strategy. All security had art training and wore badges that said Ask Me About the Art in big, bold letters. When anyone was looking confused, they were empowered to engage. They hired people with arts degrees and paid them more than you would pay a security guard and gave them training so that they could really step in and be part of someone's experience of the art, as well as asset protection.

And I wonder if your colleague visited the *Dallas Museum of Art*. I'm reminded of the *DMA Friends* program which Rob was instrumental in developing. It was a free membership program that utilised clever technology to capture data about *DMA* visitors, but also to create a more personalised experiences for a more diverse audience, in order to drive more engagement with the artwork and the institution. Rob, could you talk a little bit about the *DMA Friends* program, how it started, and what were the driving motivations behind it?

ROB: Sure, yeah. I think our observation in Dallas, like many cities in the U.S., is that there is a giant class divide between some very wealthy people in the city and much less wealthy individuals. We, of course, saw our demographics and participation follow along those lines – whether it's that more wealthy people saw themselves as consumers of art, or they have more leisure time, or that they could pay the entrance admission, whatever it was. We decided that as a museum that's ostensibly the museum for the city of Dallas, we wanted the whole city to feel like they belonged there. We recognised that there were a number of impediments, one being the admission price. Our experience flies in the face of other data that has been shared around the field concerning admission prices. Colleen Dilenschneider often shares data that would counter this,[2] but our experience was that families,

---

[2] Colleen Dilenschneider's blog *Know Your Own Bone* focusses on data-driven market strategies for non-profits. https://www.colleendilen.com/

especially lower socioeconomic status families, were able to and did come to the museum once we were able to drop our admission charges. That was an important shift, but it wasn't the only one.

The economics of a visit are paired with a harder nut to crack, which is convincing or communicating to someone that does not see themselves as an art person, that the art museum would welcome them, that they were wanted there, that they could feel smart and not stupid and not talked down to, and that they wouldn't be looked at awkwardly. Many people who've walked into art museums around the country have had experiences where they're made to feel stupid, or they don't see anyone else like themselves in the museum, or in fact they're treated poorly.

Seph, I don't know if your friend visited the *DMA* or not, but we did make a concerted effort to empower, inform and activate our gallery attendants. Early on, before I arrived, they were told that they should never engage with a visitor, that they should only protect the art. That created a very hostile environment. I think our most important change was that we recruited those team members as ambassadors for the city of Dallas and we asked them about things we could do that would make their friends and family and neighbours feel like the *DMA* is somewhere they would want to come. They were the most important shift, so that when you walked in you felt like you belonged there. And we increased attendance by 50%, and we increased minority participation by 23% in just two years.

SEPH: Can I talk about that for a minute? I love that museums in the last generation have collectively come to this conclusion that it does create, to paraphrase Rob, a more welcoming environment to have gallery attendants that are security personnel, also act as interlocutors with visitors.

Thinking about that, it reminds me of this notion that gallery attendants, once you've done the due diligence of making sure that you recruit people who are interested in the art, you may follow up with training that empowers them to engage with visitors. That's all well and good, but looking at the actual museum structure, looking at how much these people are paid versus how much curators are paid, or executives, or people who run the museum are paid. What's the incentive for that staff to be as knowledgeable as they are, potentially with art historical degrees? Can they actually afford to do that? Can they actually eek out a living? Because I think there's a way in which museums fundamentally are resistant to addressing the socioeconomic realities of being a rather badly paid worker. I see a disconnect with the rhetoric of recognising docents and gallery staff as more empowered, but not actually providing the means for them to stay long term in those staff positions. Am I wrong about this?

ROB: I think you're right in the general overarching point, but I think you are making an assumption that's problematic. The assumption that your front of house staff, the people who are primarily engaging with the visitor, should have an art historical degree or a specialisation in art history, is problematic because our research shows that the number one reason why people don't visit an art museum is that they don't think that art is for them. I would assert that it's really valuable to have people on the floor – on the front lines – who you can very easily connect with who don't necessarily have any more formal training about art than you do but have found their own point of view and perspective for why these works of art are interesting to them.

In our work in Dallas, we did not replace our existing gallery staff. The same staff that were being told to protect the art and to tell you stop touching, we kept them and engaged them in a different way. Very few of them had degrees in art history or a background in art, but many of them had an interest in art personally. I think you're 100% right – many times the front-line staff are the least well compensated staff in a museum, and sometimes unfairly so. We have an experience where we had for a long time been outsourcing those gallery attendants to a security service and they would be paid hourly and we had high turnover in those positions. It was a hard job and not well paying.

We discovered that when we hired those staff as full-time employees directly, they were more invested in us, that we could provide them better training in a better environment, and insurance benefits, and that that resulted in a better experience for guests. It's compelling in the face of so many museum expansion projects, where new wings will tally hundreds of millions of dollars. Those spaces often include large material and environmental costs that can sometimes cause museums to struggle financially to keep up with the maintenance of those spaces. It's intriguing to consider what spending a similar amount of money on endowing a living wage for staff in museums might result in.

When you think of the museum building itself as being inclusive of the staff that work there, then endowing, let's say gallery attendant positions where that gallery attendant wouldn't have to work a second job, and could afford to support themselves and a child, with good benefits, and maybe even have the ability to study and advance in their career. That would have a much more impactful benefit on the public, than an additional wing, in my opinion.

SEPH: I completely agree with you and I think that precisely at this moment, well, it's an attenuated crisis in the museum. I think what you just said needs to be said more often and loudly. There's a way in which building a new wing and expanding a new museum always gets the kinds of attention that museums ultimately want. But, to sustain that attention is an entirely different process and it's problematic. That conversation around giving museum staff the means to actually grow within an institution, starting with a living wage, is crucial.

KEIR: Having lived in Sydney and San Francisco, two of the most expensive cities in the world, where very few of the front-line staff, security staff, store staff and the people who work in the cafes, are able to earn a living wage.

Often, people up to middle management aren't provided a living wage for those sorts of cities. Beyond that, I think the attraction of the new wing or the attraction of the new building, is problematised in the U.S. where private philanthropy provides them the lion's share of resources for cultural institutions, which is not the case in the U.K., Hong Kong or Singapore, Australia or New Zealand, where government is most often the funder, and in those cases there's fewer new wings being built, because it's much easier to fund and put a name on a big new building than it is to put a name on better plumbing, better infrastructure, a living wage.

You had something that you wanted to get back to, Seph, and then I'd like to talk about the idea of a "useful museum."

SEPH: Yeah, you posed a question about whether we're at a tipping point where certain museums might be for only certain kinds of audiences. There's a pro-gun museum, an anti-gun museum. Essentially, you're talking about the Balkanisation of culture, and I think there really is that risk, at least in America, albeit not in such stark terms.

One of the ways in which we can see that risk come to the surface is in the ways that museums haven't yet developed useful ways of dealing with controversy. There are many other examples, such as what happened at *CAM*, with the Kelley Walker exhibition . . . the artist had popular cultural images of black people blown up really large on which he slathered materials like toothpaste and whatever. There was a way in which people read the work as cultural appropriation and taking advantage of people of colour, using their bodies in ways that had nothing to do with a dialogue with their culture, but rather just mining the culture as a resource for the artist. There were rather strident protests and at one point the museum actually asked the artist to have a conversation with the curator publicly, and the artist said "no." He basically disavowed responsibility for the work. He said "this is just my work and I don't have to engage in dialogue with you about it because it's essentially my own thinking and my own aesthetic production," which did not go over well.

Or at the *Boston Museum of Fine Arts*, where there was a Monet painting paired with a kimono, and they allowed people to wear a kimono and take a picture with the painting, which was tied to social media campaigns for the exhibition. That also became subject to protest.

The ways that the museums actually respond to these kinds of controversies have been disappointing because they go into panic mode. They think, "oh, well, we've got to close this down and make sure people aren't offended." I think museums should not be afraid of controversy. They should be the institutions where they grasp the nettle, and say to themselves, "yes, controversy is going to happen and we should be the people should find ways to navigate through these controversies." So that we actually end up in a place where cultural encounters aren't feared, but they're seen as moments that make learning possible, where we learn about the other, whoever that other is.

That is something that I am truly concerned about in terms of the future of museums, because if we don't learn to manage controversy in a way that is not panicked, that doesn't imagine that conversations should always be pleasant and nurturing and comforting, then museums are going to ultimately fail in the task of being the kind of institutions that can be trusted.

KEIR: That's super interesting, Seph. My thinking on this topic has been influenced by Peter Samis and Mimi Michaelson's book *Creating the Visitor-Centred Museum*, in which they write about how knowing where your museum visitors are coming from helps you engage them in a meaningful dialogue. (Samis and Michaelson 2017) Rob, those museums that choose to engage in meaningful dialogues, those that acknowledge that Balkanisation is possible and dangerous, those that choose to take a position . . . how should they behave? What tools should they be using? What would AAM advise, or what would you advise?

ROB: All the tools! I've always felt that technologies and media themselves are not inherently good or inherently evil, but in fact they're just tools that people will employ for good outcomes or not so good outcomes. What we see today is the sophistication of sales and marketing organisations who know so much about us in order to deliver and sell us products and services. We all have common experiences where that happens and it's creepy and ugly, and we probably have experiences where that knowledge allows a company to do something really delightful that we enjoy. I don't think there's any reason why museums and culture organisations in general should be afraid of using those tools to do good in the world, to create inclusive dialogues, to highlight the diversity of culture and history, to promote tolerance of different ideas and to promote acceptance of difference. Those are tools that can be very effective in shaping consumer behaviour. If we think society is in some ways similar to consumer behaviour, then I think we should go after it.

Now, I think there's some additional responsibilities of public organisations regarding privacy and control of data. But, where we can hold respect for individuals in a prime place and give individuals control of how their data is or is not used, I think we could establish a personalisation practice where we actually know what visitors respond to and we can use our agency to promote these ideals of public discourse.

KEIR: Thanks Rob.

We've been discussing transformational shifts regarding how museums create agency, allowing visitors to create their own pathways and create their own knowledge inside institutions, and how that relationship has been tested and enhanced by personalisation, sometimes using technology, sometimes using people, sometimes using marketing and sometimes using branding. I feel like this move from the privileged to the educated, and the educated to the interested, will lead in the future to a move from the interested to the engaged and participating. If you were to look 15 to 20 years out from now, what other models do you see emerging for how people will attend these sorts of memory institutions? Seph?

SEPH: I was hoping that you were going to ask Rob first! What immediately came to mind when you asked the question was two very different scenarios, one which is the old stodgy scenario – and that's admittedly not very fair – the classic institution where taking care of the object as its primary reason for being, along with the development of scholarly work around objects. I think that that institution is going to survive. In fact, I think that there are lots of reasons that we still need institutions like that and in the future, there will be lots of museums like that.

The other scenario, where the participatory drive could possibly take museums in the direction of – and I think this is a utopian thought – the direction of the *kunsthalle* where you have non-collecting institution, because honestly, the more I look at this field, the more I'm convinced that you don't actually have to collect objects in order to provide spaces for people to come together and do some thinking around them. I really believe that you can have imaginative, even difficult, encounters and conversations in a public institution around a rotating set of objects, and even curators and curatorial approaches. In the final analysis, what we want is to give people a chance to come together and have a useful friction develop, right? So that we encounter these ideas that we haven't encountered before. Why not just focus on that and let the scholarship and the care of the object fall to other institutions?

KEIR: Mm-hmm, Rob?

ROB: I think Seph is probably right that there will be a bifurcation, or at least proliferation, of different types of museums. I think we're already seeing a blending between how museums categorise themselves between history, art, science, culture, performance, in that those things are blending into each other. I think that will continue to happen. This thought that Seph raises of a museum that's centred around ideas itself, or the threat of ideas, as what the museum is about, is interesting. I could see that this could inform a collecting institution, whose collection strategy is not art historical in nature, but more philosophical or ideological. The *Lucas Museum* being built in Los Angeles is focused around storytelling and narrative, and so its collections are a forum about those concepts.

I would like to see museums being much more porous about what is contained in the museum, and what is in the community. I was in Aarhus in Denmark recently and visited their main library *Dokk1*, and in some ways it's a library that has books that we're very familiar with, but in other ways it is the public square where people hang out and use their public services, go to the motor vehicle department, pay their taxes and read books, with a great view and a playground. I think those kind of community spaces that have objects infused within them would be interesting to me. I don't know if that will be the natural future, but I think that it's a place . . . Keir, you were heading towards talking about the useful museum. I think that's a place that would be useful that doesn't sacrifice meeting what the visitor wants, or the care of objects and a provocation of ideological explorations and scholarship. I think that it's possible. I don't think we see many of those places today, so we'll have to innovate our business models to be able to afford such a place. But that's what I would hope we see in the next 15 to 20 years.

KEIR: That term, at least as I use it, comes from Tania Bruguera and her *Arte Útil* project, or as exemplified by libraries like the Queens Library in New York. They have or had a lending library of ties so that people in the community, if they need a tie for a job interview, they can borrow a tie. Which I read as coming out of the library model of being useful to the community and going beyond visitor-centred to being visitor-useful. I think some institutions will go in that direction, but it is certainly not a wave that's sweeping the sector.

Being engaged with, and relevant to, your communities in a way that they can discern and understand, and in a way that the institution's staff can believe in, is going to just as important if the museums we're building today are going to survive into tomorrow.

SEPH: Indeed.

KEIR: Thank you both, this has been great.

Seph and Rob, as Courtney Johnston notes in the preface, effectively set the scene for the rest of the chapters herein. This conversation covers recent museum history through to the present day, with an eye to the future, exploring dominant Western museum models and some of the ideas and practices that challenge them. There are two strands that we'd like to draw your attention to. The first is the discussion on trust and authority, and how trust is (or isn't) changing in response to current political, economic and technological circumstances. Does the current environment of political polarisation, arguably exacerbated by technological personalisation practices,[3] risk creating a Balkanised cultural environment, wherein people only visit museums that speak to their pre-existing view of the world, per Seph's question? How do museum personalisation practices play into or challenge these practices?

Linked to this are other questions about trust in museums, including whether we understand the nature of institutional trust; whether acknowledging the lack of neutrality in museums erodes institutional trust or helps build it; and whether museums can be trusted by each of the communities they wish to serve. One of the often-invisible communities that museums serve is their internal constituencies, such as their staff and volunteers. Seph and Rob each wonder about the impact of the sector's decisions to spend on infrastructure and new building

---

3 YouTube's recommendation engines have been shown to amplify extremist and radical messages by recommending incrementally more extremist content. (Lewis 2018)

projects over supporting living wages for staff. This line of thought connects to several major conversations happening within the museum sector about labor equity and systemic under-compensation, and the responsibilities that museums have to their staff.[4]

For more on trust, we suggest you read:

> **Conversation 2: LaToya Devezin + Barbara Makuati-Afitu**
> **Conversation 4: Sarah Brin + Adriel Luis**
> **Conversation 9: Brad Dunn + Daryl Karp**

For more on wage equity and museum labor in museums, we suggest you read:

> **Conversation 2: LaToya Devezin + Barbara Makuati-Afitu**
> **Conversation 7: Shelley Bernstein + Seb Chan**

References

Lewis, Rebecca. 2018. "Alternative Influence: Broadcasting the Reactionary Right on YouTube." New York, NY. https://datasociety.net/research/media-manipulation.

Samis, Peter, and Mimi Michaelson. 2017. *Creating the Visitor-centred Museum*. New York: Routledge, Taylor & Francis Group.

---

[4] Although the conversations about pay equity and systemic under-compensation have been continuing for some time, in May 2019, the Art + Museum Transparency Group launched the Salary Transparency Spreadsheet, which sought to make salaries within the sector visible, and to enable museum employees to discuss pay equity with greater knowledge and insight. This spreadsheet ignited a far greater and more visible public debate within the sector. It was followed in July 2019 by the Unpaid Internship Spreadsheet.

CONVERSATION 2

LaToya Devezin + Barbara Makuati-Afitu

"All stories deserve to be told."

— *LaToya Devezin*

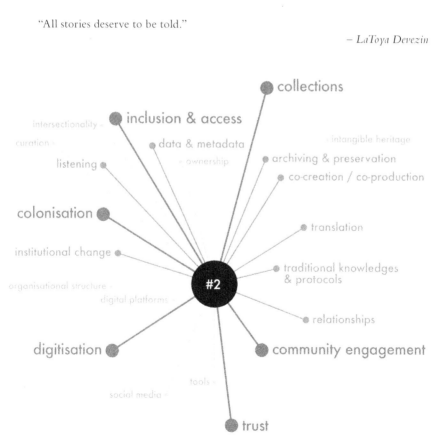

FIGURE 2.1

In 2016, Keir heard the community archivist LaToya Devezin (U.S.A) in conversation at the Cultural Heritage & Social Change Summit in New Orleans, U.S. talking about her holistic, community-centred approach to digital and non-digital archiving within communities undergoing change or stress.

A year later, Barbara Makuati-Afitu (Samoa/New Zealand) spoke at the National Digital Forum in Wellington, New Zealand about the Pacific Collections Access Project and the community-centred approach the *Tāmaki Paenga Hira Auckland War Memorial Museum* was taking with 13 Moana Pacific Island nations and groupings to document and, where appropriate, share their *taonga*.

By bringing LaToya and Barbara together in conversation we were able to investigate the parallels and distinctions between their respective internal and external strategies for archiving, enriching, and opening up cultural artefacts and stories.

This conversation was recorded on June 21, 2018. At the time, LaToya Devezin was an Archivist for *National Archives* in Atlanta, U.S. and Barbara Makuati-Afitu was a Community Engagement Navigator at the *Tāmaki Paenga Hira Auckland War Memorial Museum* in Auckland, New Zealand.

KEIR: I first encountered the Auckland Museum's Pacific Collection Access Project (PCAP) in 2017 and was inspired by how the Pacific Islander communities seem to be involved throughout the process. Could you describe the project?

BARBARA: At Auckland Museum, we have over 30,000 *taonga* or treasures in our Pacific Collection. PCAP is a three-year project where we look at, conserve, photograph, and create up-to-date and innovative ways to rehouse five-and-a-half thousand of our treasures from 13 Moana Pacific Island nations and groupings. The 13 were chosen as they represented the demographics of Auckland. Our team is made up of a conservator, cataloguers and storage technicians. We also wanted to enhance our records, and for "me" (as a Samoan), to correct and give back the voice of "our" ancestors to their *taonga*. Museums will seek "knowledge" as gospel from books about these but never from the community and the elders where they come from, which we are now trying to do.

This project has been a catalyst for change, and a platform for our communities to see what we're doing, and to start building trust with our communities so they can see this institution in a different light.

KEIR: That sounds really interesting.

When I first met LaToya, she was working at the *Austin History Center* as an African American Community Archivist. Can you tell us about the practice of doing community archiving in Austin, and the main focus of your work there, LaToya?

LATOYA: I was drawn to that project based on work I did in New Orleans, collecting the histories of those involved with the African diaspora. In Austin, I focussed on the African American community in Travis County. The members of the African American community felt they weren't receiving proper treatment at the *History Center* and weren't being acknowledged and seen. They asked the city of Austin to create a position for someone who could help archive their history.

KEIR: Could you describe the African American community in Austin? Is it a changing community? I know Austin is growing very quickly.

LATOYA: The ironic thing about Austin is it's one of the fastest-growing cities in the United States, yet it has the fastest-shrinking African American population. When I moved to Austin in 2014, the African American population was at 12%. Today, it's under 6%. That's because of gentrification and the high cost of living. People are working more than one job to make ends meet. It's a city of people who have roots in Austin, who are living pay cheque to pay cheque, just to afford to stay there.

Austin is undergoing redevelopment. Buildings are being knocked down; cemeteries are being built over. We're losing a lot of history because the landscape is changing rapidly.

KEIR: Are artefacts of place – the churches, the cemeteries, the community centres, the restaurants – part of the process of community archiving?

LATOYA: Yes. In community archives, we look at the community as a whole, from the geography of community border, to the progression of neighbourhoods, restaurants, libraries, buildings and schools.

Most early African American communities in the United States have a church and a school. If there's a historic church and school in a community, you can assume that there was most likely a neighbourhood around them. So, what's that story? Part of the challenge with community archiving is trying to uncover those narratives.

With African American history, a lot is not written down. It's passed down orally through the elders of the community. We weren't considered people; we had chattel slavery. We usually have to search slave owners to search African American history pre-1870. You'll see a tick mark rather than a person's name. You don't know where they came from or the family history. It's filling those gaps, and almost being an historical detective.

KEIR: How does that connect with what you're experiencing, Barbara? When you set out to do a community archiving project, how do you uncover those stories?

BARBARA: I am Samoan first. Who I am and where I come from has always guided how I deliver PCAP and our engagement *kaupapa* (protocols).

We would connect with community leads who are based in Auckland and started the conversation there first about what PCAP was about. That takes so much time to unpack, to build that trust, for them to understand what this place [the *Auckland Museum*] does, but also its relevance for our

Moana Oceania communities. Once the community leaders understand what we are trying to do, they would (in conversation with other leaders from their community) choose the best person living to speak to their treasures and give their voice.

We're on such a tight timeframe and budget and, as with every project, things can always be done so much better. We based how much time we would spend on each community based on the number of *taonga*, not on the size of communities. While the project plan spoke to community engagement, we didn't allow enough time as we jumped straight in. Politically, it was a minefield. In the beginning we were very focussed around co-development and co-creating this kind of relationship and what it can look like. The success of PCAP has been because the communities have come on board in their fullness, but we quickly learned that to do this properly we had to listen and allow our community leads to co-lead and only with their guidance we would be able to do this engagement properly so it's genuine. However, we [PCAP Community Navigator team] had to prepare the *fale* [Museum] for our communities, we had to navigate and ensure the *fale* was ready for them culturally, as well as spiritually and physically.

We are engaging with 13 Moana Oceania communities – all with very different and unique cultures and traditions – so we needed to build 13 different *va* [relationships]. We needed to learn and understand, and be open to learning, their cultural protocols and traditions and how we could ensure we opened up our *fale*, and not trying to get them to fit into a *palagi* Western kind of way.

KEIR: I've heard the phrase, "nothing about us without us," which resonates with me, but it's not as simple as that. It's also, "don't begin the process without understanding that our processes exist already." It's not just including people in the storymaking, it's including people who are the subjects and the experts in the process of developing the plan to start generating and creating the stories. That's something our sector struggles with.

LATOYA: I agree with you.

BARBARA: Yeah, me too.

LATOYA: That's the difference between community engagement versus co-creation. We ask the community to come to us and we put a lot of onus on the community. We tax them quite heavily, without returning that investment.

BARBARA: Absolute, yeah! I love you, LaToya.

KEIR: The part that I'm not seeing anyone do well is coherently and respectfully incorporate co-developed and co-created knowledge and information into the archive. What are your thoughts on effective strategies of taking co-created and co-developed content, stories, resources and materials, and bringing them into the institutional framework? What approaches have you found effective?

BARBARA: Part of the [Pacific Collection Access] Project was to enhance our records. The Indigenous voice is being given back, so when people around world look for, say, a fan from the Cook Island, they'll be able to look it up under *ili*, using their own Indigenous name.

Everyone here feels that they are guardians, we call them *kaitiaki*. The collections aren't anything without people. When I first heard about this project, I assumed that the institute [museum] understood what it meant to culturally and correctly host the different communities they were inviting in, but none of that was set up. There was huge internal navigating to do, with quite a few challenges to shift that mindset.

The creation of my role challenged many in the building as to A) opening up the collection directly to our communities, and B) what it should look like from a museum perspective. But mindset takes years; it doesn't happen automatically. So, having to lead by example as to what it could look like when we have our communities in the museum.

I had to create general Moana Pacific protocols around the appropriateness of how to host, but this changed quickly, as we had to learn to tailor on the move. I could only speak as a Samoan, and through the initial engagement I would often say that I am Samoan and that's what I know, but that I needed them to guide this *va* [relationship]. I realise the importance of our roles, because if we're not here to try and push it as Indigenous, then things won't change.

Our Fijian community, when asked what would have helped make this relationship smoother during our evaluation, suggested *Tikanga* 101, or a Protocols 101 led by our community leads. This has now become the norm at the start of each engagement.

KEIR: I like the idea of doing these full-circle evaluations between each step and improving the process through the different communities. You're lucky to have a number of steps, so you can inform your practice going forward.

How does that compare to your experience in Austin, LaToya?

LATOYA: We're doing a lot of the same practices, with the opening of the archives and letting people see that process and what it looks like, and being open and receptive to change. With collaborative archives the vocabulary and terms that we use are important, because language is so powerful. You want to make sure you have the correct terminology.

With the controlled vocabulary, most libraries go by the Library of Congress subject heading. Some headings have to be changed, some of the vocabulary has to be changed – like what Barbara says about using a tribal name. For instance, instead of using the term Latino at the *History Center*, we use Latinx. We no longer use the term Hispanic because of its colonial connotations. The community in Austin felt more comfortable with Latinx and that's gaining popularity as the preferred term for Latin American populations.

KEIR: Did you fund and resource specialists for specific communities? Did you have an African American Community Archivist and a Latinx Community Archivist?

LATOYA: Each community archivist at the *History Center* worked with different communities. I worked with the African American community, and I'm of African American descent. We had a woman of Chinese descent who worked with the Asian American community, with both the South Asian and East Asian communities. We had another archivist who worked with the Mexican American and Latinx communities.

Before I left, my suggestion was to change the term African American to just black, because it's more inclusive. For example, the Nigerian population is growing in Austin, and that's a community that's not being served as much.

KEIR: What sort of reactions do you hear from members of the communities you've worked with? What are they hoping to achieve when they partner with an organisation to give their stories? To give away physical objects as well as memories? Do you have to convince them to partner with an archive, or are people already looking for the right institution to connect with?

LATOYA: It varies. As an African American woman, I'm one person and I can't speak for all African Americans. It's the same thing with collecting the histories. Some people are willing to give their history to you, because they want to preserve it, and some individuals are not comfortable because they've been hurt by different systems that are in place when it comes to collecting and appropriating the stories. That's a big issue that we experience with the African American community in the United States. We take and take, and take, and take, but sometimes don't give back to the community. Even though we know the history, we know the stories, the community sometimes still isn't supported.

That's something that we're working actively to change.

KEIR: LaToya, I know that on the east side of Austin, development and rising property costs continue to force African American communities out. Is that something that affected you personally?

LATOYA: Yes. Even though I loved my job, in January I had to relocate to Atlanta, because I could no longer afford to live in Austin. I was working two jobs, and I needed a third one. As the city's costs were increasing, our salaries were not, and it was difficult to stay. I had so many projects that I loved and was involved in, and I'm still working on. It's about continuing those narratives and those stories. It's something I feel like I have a lifelong commitment to, so I will never stop working with community archives.

KEIR: That's real.

Barbara, I would love for you to describe or explain the concept of *teu le vā*. My understanding is that it's a way of creating a culturally, spiritually and digitally safe place. I'm not sure if that's correct. Could you explain the concept and how it applies in the PCAP project?

BARBARA: *Teu le vā*[1] is our Pacific strategy that was created by the Pacific staff about five years ago in response to some big changes that were happening in

---

1 The *Teu le vā* guiding principles articulate an operational and philosophical culture for the museum based on Pacific perspectives and cultural values that make the museum a respectful, relevant and

the museum, recognizing that we're not Indigenous to this land and proudly Moana Oceania (Pacific). *Teu le vā* is a Samoan proverb about the sacredness of relationships. It's that sacred relationship that you have with each other, but also with the environment, with the gods, and with the *fanua* . . . the land.

The principals that the staff had created are, for me, the founding document to how we're trying to deliver the Pacific Collection Access Project. It's also been a vehicle to educate others, so that they understand exactly what we're trying to do as Moana Oceania (Pacific), in this time and in this place.

KEIR: Does that idea extend into the digital space? Or, is it about how people engage with physical objects in the museum and physical objects in the museum's archives?

BARBARA: I feel in conflict of myself every day working here, which is not a good thing.

Let me backtrack . . . With the knowledge holders that we were gifted through the community, we had to go through quite a difficult internal process – which had a lot of tears, as we as Moana Oceania have a completely different understanding to what this relationship means – but to get this institute to recognise what these experts are in their fullness. These knowledge holders are makers and descendants of those that have carried knowledge – sacred knowledge which has been generations carried. They are respected elders in their respective communities, some hold chiefly titles, some have been sharing this knowledge for years and tailored to what fibres Aotearoa has to offer. For many years this Indigenous knowledge was never acknowledged for expertise in this sector, so I am grateful the museum has set a precedent for others to come. We set up a proper contract with each of them so that they could get paid for that knowledge, which is respectful and right.

---

engaging place for Pacific people, and reflects the stories of Pacific people and cultures. Though Pacific communities are vastly diverse, commonalities can be identified.

- Meaningful engagement – develop and sustain relationships with Pacific communities
- Integrity – once established, maintain principled relationships
- Authenticity – Pacific voices, languages, and perspectives are included and valued
- Reciprocity – acknowledge rights of source communities and the museum's responsibilities by actively building capacity and enhancing capability
- Responsive – empowering source communities to achieve their aspirations
- Balance – recognise mutuality of power, control and involvement
- Symmetry – recognise that communities are knowledge holders of items within the collections
- Enact obligations – demonstrate value of significant Pacific collections by resourcing the care and storage needs, and building knowledge of those collections
- Respect – recognise that contextual information is held within source communities
- Value diversity – recognise uniqueness of Pacific cultures, their cultural frameworks and stories; facilitate the pre-eminence of Pacific languages pertaining to Pacific collections on-site, off-site and online

I get challenged about putting a price on this oral history that has belonged to us and been carried by our ancestors for generations, and which is very sacred to us. Yet here we are putting a price on it so that it can go into these archives of an institution that isn't ours. This is why we have to be in these spaces – to challenge.

Originally, when the knowledge holders were first coming in, we had to create a safe platform for them to have a voice in this kind of institution – this includes *tikanga* (protocol led by them) and building trust. Our knowledge holders were grateful to be the ones chosen to give that voice for our ancestors and for us to be acknowledging, through protocol, it is right and needs to be embedded as the norm in the GLAM sector.

In saying that, there is some criticism around how much [knowledge] is being taken. That has been one of the things that I've learned, after the fourth or fifth Moana Pacific community that we've worked with. We would ask our knowledge holders to share for example on, "what was this bottle called," "how was it made" and "what were the materials used." But, when some of our elders have come (and we've – the museum – created a truly safe space), they would open up and share so much including sacred knowledge. We, as the museum, don't have the right to hold much of this knowledge as it's not ours and we need to be brave enough to voice that – as Indigenous – that to name/identify certain knowledge would enhance our records but this sacred knowledge and parts of it are not the museum's to hold, but that we can help build a framework that will enable our communities to hold and control their own.

We had a Kiribati knowledge holder session once where our knowledge holders kept referring and speaking about "Kiribati magic." One of our Kiribati knowledge holders stopped the session and said it perfectly, "Museum staff please note that Kiribati magic is not for your records/labels as its specifically for Kiribati only and without anyone here who is Kiribati to speak to it. It is not yours to note." Museums need to learn to build Indigenous permissioning for our Indigenous knowledge that gives them full voice and power for their own. We need to understand that these Western frameworks are often not relevant or appropriate in many cases to suit us.

We also need to start looking at "How do we create a framework for our communities to have it, and to hold it, in digital space?" We enhance our records based on these kinds of projects and quickly look to share online, but we have not ensured we have our knowledge holders' final permission embedded in the policy and process to ensure they have final signoff before anything gets finalised.

KEIR: It's something I've been grappling with, these library catalogue metaphors that have extended into museum collection management, where you have things like binary genders. You've got to check a box where you can make an artist a male or a female, however we have artists that don't identify as male or female in our collection. Or where there is no single creator, where it's a community creation. It's almost like the record doesn't want to be saved when you haven't met these data standards as inherited into a system that doesn't have the nuance to create a meaningful record of what you're hearing from the creator or from the community – the storytelling around it. The digital tools to encode the meaning around our collection objects are often oppressive. They exclude people just by the way that they're architected, and that's a foundational problem.

BARBARA: I'm nodding my head.

KEIR: There's a kind of false premise that people should willingly give away their stories, their knowledge and their time for free.

I don't know if you know of the project, *Philadelphia Assembled* by Dutch artist Jeanne van Heeswijk, presented at the *Philadelphia Museum of Art*. It was a city-wide set of engagements that culminated in an exhibition built, largely, on the visions and hopes the people of Philadelphia have for their city.

For me, the most radical act is that she paid everyone involved in the project the same amount for the hours that they contributed. Everyone, including the artist, got an hourly rate, whether they were a superstar designer or someone who had a unique story.

Museums and archives frequently ask people to give up their time and energy with no remuneration, with the threat of, "if you don't, then we won't be able to capture these stories and they will disappear." There's a perpetuation of colonial behaviours by the GLAM sector that are dangerous, and aren't being talked about, or at least not in the forums that I am in. Is that something that you feel on the ground?

LATOYA: That was one of the biggest complaints we had in Austin. One of the developers who was responsible for the displacement of communities would have meetings in the middle of the day, and some of the women and parents who were in attendance were saying, "You have the meetings in the middle of the day, my children are in school, I have a job, I'm working two or three jobs, what about Sundays?" Most of the people there asked if the developer could meet them at church or come where they are. The meetings would be in obscure places where it was difficult for the community to reach.

That's the problem with community engagement. Sometimes we expect people to come to us, but we have to go to them. That was one of the reasons we founded the community archivist program, to embed us into the community so that we were accessible.

When we make documentaries about people's stories, that's a big area for people feeling robbed – becoming victims of cultural appropriation. Filmmakers come in and do research for a month, meet with members of the community and make a documentary that might be a blockbuster, but the people who participated were not compensated. That happens quite a bit where the creators are receiving financial compensation, but the community is not.

The Philadelphia project that you spoke of, I think that's great that everyone was compensated. We did that with a project that I worked with called *Invisible Intersections*. It was an art exhibit focusing on invisibility and displacement of communities of colour in Austin. We made sure that we paid both the subjects and the artists. It was difficult to get the funding for it, but we did.

KEIR: Yeah, I think you've got to fight, don't you? Barbara, does that resonate with you and your experience with PCAP?

BARBARA: Yes, absolutely. When the museum had originally done PCAP's project plan, before me coming on board, I don't think they had budgeted enough for community participation. They assumed that the community would just come and give this sacred knowledge that had been carried for generations, for free. Or, we would give a *koha*, which is like a donation.

Once you open that door, you need to be ready to do whatever you have to do. Our communities don't RSVP, so we've had groups who've said there will be 20 people coming, and yet they'll arrive with 250. In our Moana Pacific cultures and protocols, we have to break bread together. What does that look like in the institutional setting? It's also finding ways to ensure that it's recognised that this is how I, as a Samoan, as a Pacific person, have to do things culturally, and looking at ways to bring everyone along the journey of understanding. It's about educating and recognizing for us, as Pacific, this isn't our *whenua*, we are NOT Indigenous to this land, and how through protocol (theirs and ours) that we acknowledge *tangata whenua* . . . the people of this land with respect.

When we were expecting the 20 and had 250 turn up, I was freaking out and ringing my family to say, "Can you please order 50 pizzas and get them delivered?" and they were all laughing on the phone saying, "You must be crazy." One of the elders tapped me on the shoulder and said, "We brought food." They came in with chilly bins both full of chicken, potato and fish. Once you open the door and host well, it lays a foundation of such amazingness. You then have to keep ensuring that the relationship is nurtured and held strong throughout, even once PCAP's gone.

KEIR: Should resourcing the relationship into the future be a key part of designing the project?

BARBARA: Yes.

LATOYA: There's another statement we hear from the community. They say, "You come to us and ask us for our histories, our stories, our objects, our narrative and then you don't follow up with us." We don't nurture that relationship. We go in, get our stuff and leave. In Austin, I maintain relationships with a lot of the members of the community even though I'm not attached to the institution anymore. That doesn't mean that I can't foster those relationships, because I care about what happens.

That's one of those things – making sure that you continue that relationship. Going to the events the community is hosting, showing up sometimes and not saying anything. If there's a dance, go to the dance. Get to know the community, witness the community and see what they're doing and what they're about. Sometimes we go in like we have all the answers, and we talk over the community instead of listening and learning about their experience. With gentrification especially, we try to change these neighbourhoods instead of learning about what makes these neighbourhoods great and saying, "If there is some need for improvement, how can we do that together?"

KEIR: How do you define and measure the success of a community archiving or preservation project? For PCAP, what are the established measures you're going to use when you think about whether or not it's been successful, Barbara?

BARBARA: My role is to ensure that our communities are taken care of when they engage with this place, but ensuring that as a museum, we have real tangibles, like protocols in place around dealing with the different communities. A lot of our *teu le vā* principles and practices are embedded throughout the organisation, and also in policy. But, first and foremost, it's the communities being happy as I came to serve my community first.

I hear what you're saying, LaToya. You have to go to all of these community events, and you have to be present, so they can see that you are there to support them. It puts a huge onus on us as the ones that hold that relationship. Someone recently asked me, "What happens when you leave? What happens when this project finishes?"

The big success for me would be that other museums adopt this kind of practice and don't see it as something scary. The reason I go to all of these conferences is that I'm hoping that these kinds of practices are seen positively, and that other museums or organisations in the GLAM sector will be brave enough to know that this is the right way of doing it.

KEIR: This sort of work often gets hidden by the big, shiny, new digital things, so it's fantastic to be able to talk to you and share your stories.

LaToya, is it important for you in your current role, and your previous roles, to see a change internally and in other organisations, or do you really focus on the community stories being captured in a co-creative and community-led way?

LATOYA: Archivists always look first at enduring value, preserving this history, this legacy, this narrative. But another movement that's happening with archives is the post-custodial model. What we're doing is trying to flip that narrative, where it's more important to preserve a community's culture and the story. Looking at people, places and narrative. It's this post-custodial model of not having to physically house everything. That's more important than saying, "I own this photo of this family or I own this object." Some of the history can stay with the community.

I still feel complicit. In New Orleans [after Hurricane Katrina] I saw so much history that was in people's homes wash away. In archives, we have a process called LOCKSS, which means *Lots of Copies, Keeps Stuff Safe*. By having multiple copies of these narratives, we hope that we won't lose anything. So that, just because an item is gone, it doesn't mean that the history is gone.

With digitisation, there's privilege with it too, because we have communities that don't have access, and there's still a digital divide that we aren't addressing.

KEIR: Technology can expand the number of people that have access to something that could otherwise be unavailable to them if they can't physically visit. People who wouldn't traditionally attend an archive, a history centre or a museum can discover things through links, friends and online communities. But, given what you've said about making sure that people can maintain ownership and maintain access, does the reliance on technology and online access unintentionally discriminate against those across the digital divide or older generations? Or is there still value, because by digitising and by syndicating, we're able to reach diasporic communities, or people who aren't seeing themselves in official histories and narratives? It's a digital dilemma. Are all of those risks and dangers worth it because the overall outcome is for the better?

LATOYA: It's a loaded issue.

Social media connects us, and with different platforms, you can hit different audiences. At the *History Center*, for instance, the Facebook group was a little older, our Instagram audience was younger. Snapchat, which we were looking into, was even younger. It's engaging with multiple platforms with different groups, and linking that data and story together, and the commonalities that we all share and what makes us different.

Social media does help engage, but in the U.S., we also have a lot of literacy issues. Some of our education systems are not the best. We have children who aren't learning, and who aren't getting the opportunity to use the internet and interact with computers, even though they're so much a part of daily life for most of us.

I worked with the Internet and American Life Project[2] a few years ago. We found that people of colour in the United States used their cell phones more to access the internet and digital collections. If we're thinking about digitising something, can we use it in different electronic platforms? Can you use it on a desktop, on a cell phone, on a tablet or on different devices? That's an important thing to consider when we look at access to history. The technology can help give us a sense of togetherness, but it can also divide us. It's a complicated history with technology, privilege and access.

Working in the public library in New Orleans, we had several customers who would come in who were not able to read. It's difficult but not

---

2 The Internet & America Life is a project of the Pew Research Center, a nonpartisan organisation in the U.S., which produces reports exploring the impact of the internet on families, communities, work and home, daily life, education, health care, and civic and political life.

impossible to work with a customer who can't read, and therefore can't get a job that would require this skill. There're so many limits that happen as a result of not having that foundation. Those are things I keep with me as I approach this work, because not all communities have access to the internet.

BARBARA: One of the things I'm mindful about, when working with 13 different communities and only being from one of them, is what should be shared and what shouldn't. But it's funny, because I see all of these elders with mobiles and tablets, taking photos and sharing it with all of their families and their communities. I'm so wary about sharing online. Some of our treasures have human remains, and I would never share those. For them (our community) it's so unique and important that they have to share it, but for the organisation, it's not okay. We need to get that conversation right, that protocol for that space *together* with that particular community.

KEIR: Is that because these communities are diaspora communities and they're largely connected online? They're spread out across the region and across the world, and this is a way to reach the whole village?

BARBARA: It's reaching the most remote villagers. . . we went to Savai'i, one of the most remove villages, and the people were like, "What's your Facebook page? What's your Facebook name?" Even taking photos and sharing them without access to Wi-Fi.

There's a huge desperate need, especially for our second and third generation born in New Zealand, Australians – abroad. They want to know so much about their culture, but they won't be able to ever get to the museum. Creating a digital space for them to see their treasures and see them by their own Indigenous name, especially for those who don't speak the language, is an amazing platform. But I am always wary about what we share, because it's not mine to share.

KEIR: I remember reading someone's reaction to PCAP was, "I never thought the museum had anything relevant to me or my family. But now I have seen this, I will bring my grandchildren to visit." When you think about archives, how important is it to be connecting the different generations?

LATOYA: It's really important to have those intergenerational connections.

One of my favourite projects is to work with 13-year-olds and partner them with senior citizens so they can learn about what life was like when the seniors were 13. Sometimes they're surprised because they see themselves in the archive, because they never think that they would see themselves reflected in our collection.

Communities of colour in the United States don't always look at the archive as a place that includes them. They are institutions that preserve a narrative, but it's often a one-sided conversation about history, and it's not theirs. When I first started at the *History Center*, there was an exhibit on the Civil War, and it talked about slavery and states' rights, but it did not include the Mexican story at all, or stories about Native Americans who were in the area. We realised that we didn't have any information about what Mexican American life was like during the Civil War, because Texas was Mexico before it was part of the United States. It's things like that that we miss sometimes.

My favourite story is when we had a family walk into the *History Center*, and there was a picture of a cheerleader on the wall from the 1940s, cheering on a football team. The grandmother saw that it was her high school picture. This was a huge picture, and it was just her as a teenager. The father made a joke and said, "Oh, look, your grandma's so old that she's in a museum." She never thought that she would see herself like that, on an 8-by-10-foot poster.

KEIR: That's fantastic.

BARBARA: That's cool.

KEIR: Barbara, when doing these sorts of community-based archiving projects, how important is it to be thinking about it across the generations from the outset? Is that something you see as a goal?

BARBARA: Oh, absolutely. In the beginning, it was a real push for us to get people to participate in PCAP, because Pacific and Maori have very low visitation to the museum, but it was because it was never a place our communities saw as "theirs." We started with the Cook Islands, and I am lovingly very grateful for the lead and knowledge holders we were gifted. The Cook Island community set an amazing bar, but it was very much based around *whanau* – around families. We would have three generations of one family come in, looking at treasures that they never knew were here or seen in "their lifetime," but heard about at primary school. Then we would hear the conversation of grandpa talking to his son and to his granddaughter about that particular treasure, because he had seen his mother making it back in the islands.

It's creating that safe space for them to come – based around tikanga (protocol). Our community days would not only give our younger generation an opportunity to see "their own" treasures, but often they would ask to be given the opportunity to do a performance (honouring this visit/activation through appreciation and performing their tradition dance), and for them to perform in this kind of institute was such a buzz, especially after just seeing their treasures. But then, to be able to see their treasures, their *taonga*, that one of their ancestors made, has taken it to that next level.

Having repeat visitors, especially the elders, coming and bring their families after the visit to PCAP, that's been the joy to see them here in this space and taking pride and ownership in its relevance to them.

KEIR: How can archives respect and protect the sacredness of these shared stories and materials, but do so in a sustainable way? How do you encode that personal, nuanced ask from the person providing that material, then bring that into the digital safely? Is that even possible?

BARBARA: I feel like it is possible when you've got people like LaToya in this space. It's someone fully understanding, culturally and spiritually, the sacredness of that knowledge that isn't ours, that isn't mine, and wanting to know how we can help – using different platforms – to ensure its safety. Just hearing you, LaToya, gave me hope, because that has been a real heaviness for me. People keep saying, "So, what are you going to do?" And I'm going, "Just giving it to Jesus, and I'm just waiting for the answer, because I don't know."

This isn't my sector. My whole thing is working for the community, and that they are at the centre of everything I do. Hearing you has given me a bit of hope of being braver to start the conversation around what we can do and how we need to shift.

LATOYA: I've enjoyed seeing the commonalities that we share even though we're on opposite sides of the world. I know for community archives, when we look at what we can put online and what we can't, sometimes communities will give you things, but they may say, "I want to wait 30 years before we release this information." If you go to local history archives, and sometimes state archives, too, you'll see records that are sealed, where they're not to be opened until a certain time. That's one way.

Getting permission from the community before you can use the records. I think the answer lies in collaborative partnerships either between institutions and the communities, or on a larger scale, like a national digital initiative. Having those partnerships in place helps protect the community. If, for instance, a filmmaker cannot use the records without talking to the community first about how they may be used, and to make sure that a story is accurate and reflective of the true narrative, that's important.

With the federal government, you can't put anything with personally identifiable information. For instance, Social Security numbers. Those shouldn't be open to the public for a living person. The archival rule is usually that the dead don't have a right to privacy, but the living do. Sometimes I walk that fine line and use the community to guide me. You must think about how the materials are being used. All stories deserve to be told.

Recently, at the *History Center*, there was a collection for the first school shooter in the United States. It was a university student at the University of Texas in Austin. The *History Center* changed that collection's name to not give the shooter glory, or to preserve the shooter in perpetuity, but to preserve the moment and what happened, the narratives of those who survived and those who passed on. It's to preserve the community's stories instead of the shooter's. His name is in the collection, but this way he's not being given glory to hopefully dissuade others from going in that direction.

BARBARA: One of my conflicts here in this museum has been around cultural permissioning for us using different images and sharing the treasures. I naively, again, assumed that when the museum was talking about cultural permissioning and our process, especially around the Pacific treasures and the Pacific knowledge, that it was done in partnership with the communities. When I challenged that, it was very much. . . "But then it becomes too hard." But who is it hard for? If you're going to do this genuinely, then you have to do it with community. Obviously this comes with its own challenges, but to start the *korero* (conversation) we will be able to come with a better process that honours and includes our communities' voices and knowledge.

KEIR: It is a delicate process of identifying people with knowledge, stories and agency, asking permission, gaining trust and respect, listening to their stories, creating real human connections, being grateful and trying to co-create something with a community that is ongoing, in ways that enrich the archives, and also enrich the communities and their participants, and making some of that knowledge more broadly and digitally available to the public where appropriate. It almost feels like this is a sacred act.

What might the museum world be like if the practices that you and those like you are pioneering were adopted by everyone? What might archive practice look like in, say, 20 years from now, if this was how everyone did it? What would that future look like?

BARBARA: It would look beautiful. Like LaToya said, with anything with community engagement, it has to be reciprocal. To be able to say that we started this conversation, and that genuine change was embedded throughout the organisations and their practices in everything including policy throughout the sector would be an amazing thing. It's a no brainer for me.

For me, it's always been it's the right thing to do. I often say that if I don't get a punch in my stomach after saying or doing something, then I've done something right. Recently, I've felt a lot of punches in the stomach, and I think that's because everything is happening quickly – there is a push to get these projects done based on budgets that were created without any flexibility. But to be brave enough to say that these are the things that need to change, that we need to be open and more fluid, and the reasons why, and not just knowing that it's the right thing to do. . . . Leading by community, with community and for the community.

LATOYA: We're going to have a more multidimensional viewpoint of history, and I think we can really affect social change. We'll have more diversity within our staffs and a more enriching experience as a whole. I feel like our space will be more transformative over time, and that we can grow with each other, and maybe it will give us the space to have some of those difficult conversations to be able to foster change. If we all could kind of participate in conversations like the conversation that we're having now, I can see it being such a beautiful and wonderful process.

KEIR: Thank you both so much, I want to end on that note, LaToya. That is a really bold vision that I would love to contribute to. I hope our conversation here today is a tiny, tiny piece in support of it.

Massive thanks to you both, this has been really wonderful.

Although co-creation is becoming a foundational competency in museums, Barbara and LaToya complicate this practice by questioning if museums are appropriately prepared for the kinds of ongoing work necessarily required to give communities agency over their objects, stories and histories – especially in digital platforms. Together, they show how using protocols and practices that are developed with and for source communities lead to better outcomes, richer information (data and metadata) and, in many cases, internal organisational change. Trust, consent and agency are core to these processes, as is the building and maintaining of long-term and mutually beneficial – rather than transactional – relationships that center on the needs and wishes of the source communities.

A particular challenge for those thinking about technology in museums is collection documentation practices and systems, which embed colonial and Western biases in ways that can minimise Indigenous knowledge and make it difficult to encode metadata for First Nations objects in a culturally grounded and appropriate way. Kimberly Christen et al. (2017) have described the unmet needs of commercial collections management systems as including, "cultural protocol driven metadata fields, differential user access based on cultural and social relationships, and functionality to include layered narratives at the item level." There are linked concerns about Indigenous data sovereignty, or the right for Indigenous communities to collect, govern and own their own data, and data related to their objects. We encourage those wishing to learn more about these issues to explore the work surrounding the open-source collections management software Mukurtu CMS and the related Traditional Knowledge (TK) Labels. Both projects demonstrate the benefit of long-term engagement with communities to solve specific digital cultural challenges.

For more on equitable, consent-based work with communities, we suggest you read:

> **Conversation 4: Sarah Brin + Adriel Luis**
> **Conversation 11: Tony Butler + Lori Fogarty**
> **Conversation 12: Daniel Glaser + Takashi Kudo**

For more on new and existing forms of colonisation, including the technical systems and established practices of digitisation, we suggest you read:

> **Conversation 4: Sarah Brin + Adriel Luis**
> **Conversation 5: Sarah Kenderdine + Merete Sanderhoff**

Reference

Christen, Kimberly, Alex Merrill, and Michael Wynne. 2017. "A Community of Relations: Mukurtu Hubs and Spokes." *D-Lib Magazine* 23 (5/6). https://doi.org/10.1045/may2017-christen.

CONVERSATION 3

Lara Day + David Smith

"How do you respect the specificity of all the different micro- or sub-cultures that exist within Asia, and make it part of a broader history of visual culture"

– *Lara Day*

FIGURE 3.1

M+ is a new museum for visual culture in Hong Kong opening its doors in 2020 with a focus on "visual art, design and architecture, moving image, and Hong Kong visual culture" (M+ 2018). We wanted to talk to Lara Day (Hong Kong) before the museum opened, while many of its initiatives were happening online or as pop-ups. Prior to joining *M+,* Lara worked in publishing with experience across books, magazines, newspapers and online, including time at the *Wall Street Journal*, *Travel + Leisure*, and *Thames & Hudson*.

Colleagues and peers in collections and registration departments, and those working in archives and libraries, encouraged us to investigate the *Asia Art Archive* (AAA). The *AAA*, situated in Hong Kong, is a self-described "catalyst for new ideas" through the "collection, creation, and sharing of knowledge around recent art in Asia" (Asia Art Archive 2017). David Smith (New Zealand/Hong Kong) has over 15 years of experience working in the audiovisual and cultural heritage sectors, and joined *AAA* from Europeana, a European metadata aggregator that enables and promotes access to cultural collections. Prior to Europeana, he worked at the *Imperial War Museum* in London, U.K. and at *Archives New Zealand*.

M+ and *AAA* are very different organisations in terms of scale, collection, audience, remit and history, however, they are both relatively young multilingual and multi-modal organisations situated in a highly mediated society with a local context and a continental remit.

This conversation was recorded on February 14, 2018. At the time, Lara Day was Manager, Digital Programme, *M+*, Hong Kong and David Smith was Head of Collections and Digital Experience, *Asia Art Archive*, Hong Kong.

KEIR: Lara, *M+* is scheduled to open in 2020 as one of the largest museums of modern and contemporary visual culture in the world, covering work from Hong Kong, Asia and beyond. The museum will be accessible through a multitude of entry points, with offers like exhibition spaces, cinemas, performance, café, lectures, archives and a study centre, media technology, roof terrace, and member facilities and restaurants. It seems like an "all things to all people" offer.

I'm interested in how cultural spaces represent institutional identities and community identities, and, in some cases, national identities. How are you describing *M+* internally? Where are you on that trajectory? How are you defining yourselves as an organisation?

LARA: As you've just highlighted, we're not just a Hong Kong museum, we're a global museum that has a Hong Kong perspective.

We are building something that did not exist before. When the museum was established in 2012, there was no collection. There was no identity. There were very few staff. Our founding executive director was the first staff member. A year ago, there were about 40 staff. Now we're at almost 100 staff. By the time we open, there will be 300+ staff. Imagine what this does to the day-to-day dynamics of an institution. We started off as something like a start-up. By the time we open, we'll have existed for eight or nine years, at which point we're an institution.

It's really important for *M+* to be part of the fabric of Hong Kong, and not seen as this giant alien spaceship that has just landed. We want to establish ourselves as a museum that people across the world can look to, but also to really be a part of Hong Kong.

Part of that is collaborating with organisations like *AAA*. David and I are planning to organise a Wikipedia editathon on women in fine arts and practitioners in Asia to align with the global Art+Feminism movement. There is a lot of work that we need to do institutionally to correct the gender imbalances in the art world, especially in the Hong Kong art scene.

DAVID: Yes.

KEIR: The way I understand it, *AAA* is framing itself is as a catalyst for new ideas. Those ideas manifest through building tools and communities that collectively expand knowledge around recent art in Asia. Can you explain that strategy?

DAVID: *AAA* was founded in 2000 to address a perceived lack of easily accessible primary and secondary source material related to the development of contemporary art of Asia.

There are three layers to our work. We've been building our physical and digital collections, including ephemeral material like exhibition catalogues and monographs, for 16 years. We have a team of researchers who are looking at areas that are less well represented in the wider narratives of the art scene and art history. Then we digitise those collections and bring those to *AAA*.

Across parts of Southeast Asia, there isn't the level of art infrastructure that there is in the West. There aren't art museums. There are fewer commercial galleries. There aren't art schools and universities. Instead, artists are self-organising through, for example, artist-run spaces. We look for where artist-run spaces have happened, and we talk to them and develop those collections.

Similarly, in Southeast Asia, performance art is a vibrant artistic practice. People create performance art festivals. We work with local practitioners who are documenting these festivals and performances, then we digitise them, make them available on our website and activate those collections through research.

On top of that, we have education programs, or what we'd call "activation."

We're totally independent from government. We get our funding through grants and philanthropy.

KEIR: The best contemporary art usually happens in the margins. It sounds like that's where you're working.

Meanwhile, *M+* has described creating a hub of activities that are sometimes artistic and sometimes not. Both organisations describe a strategy that I read as community building. What are your goals there? How does digital help? And how do the physical and digital communities mesh?

DAVID: We went through a big organisational restructure a couple of years ago, as we'd been doing all of this work, but not thinking about who our key users were. We were still thinking, we're doing this for the greater good of humanity – you know, this amorphous, non-defined group of people. We learned that our key audiences are academics and university students, curators and art professionals, and the education sector. We've stopped engaging with school kids directly. A lot of our focus now goes into teachers, with the knowledge that it will spread out to the students.

LARA: At *M+*, we talk about the museum as being more than a building: it's the relationship between our content and our audiences. Interestingly, digital here is situated in the curatorial team, which you could see as a "content team" comprising curatorial, learning and interpretation, digital and editorial, all reporting to the Chief Curator. While our building is under construction, digital plays an especially important role in bringing our content out into the world and building communities in the places that people inhabit online, like social media. We've also been experimenting with how to bridge digital and physical through the *M+ Pavilion*, a gallery space next to our construction site where we've been staging exhibitions before the building opens – for instance, giving exhibition visitors a chance to ask questions to curators via an "Ask the Curator" series published on our blog.

KEIR: It's rare in arts organisations to take something as foundational as your audience and rethink it. In *M+*'s case, the institution seems quite confident about who the audience is. Is that fair?

LARA: It's fascinating that's what the perception is. What's currently clear is the "business to business" relationship between our institution and other organisations. We need partners and strategic partnerships. We need to be part of global museum culture.

I don't think we'll know exactly who are audiences are or how they'll end up using our building – this vast museum space in the middle of a cultural district – until the doors are open. We have to be tactical and creative about audience research, and the segmentation and definition of audiences. How can we create a visitor experience where everyone feels welcome? We're looking for ways to work together across the institution to ensure that we're serving audiences across our different spaces, online and offline. That's the "everything to everyone" ambition that we have.

DAVID: It's interesting to consider getting that balance between the big rock star exhibitions, which might be predominately Western, and exhibitions that you want to challenge the audience with.

Because it is a visual culture museum, you have to be careful to not fall into it being a contemporary art museum. So, for instance, architecture is important.

LARA: That's right. The framing is something that we talk about a lot. What is visual culture? My hope is that it's going to be an entry point for a lot of audiences – because we collect things like neon signs, and archives of building projects that people might already be familiar with. Although the design and architecture objects in our collections are a great way into what we're trying to explore, when you talk about "visual culture" and no one's heard that term, that's something that we have to grapple with.

It gives us an opportunity to reframe what a contemporary museum is doing.

KEIR: Definitions of visual culture can be very . . . very broad.

DAVID: A nice piece of artwork I saw a while ago took the Kowloon Walled City fight sequence from the first *Call of Duty Black Ops* game. The artist used the tone free rendering of it and created videos without any machine guns, just looking at these contemplative spaces that the developers had created. All of that development work is done in the U.S. None of it was really developed in Hong Kong.

Hong Kong has an amazing video game history, but there was very little software development in Hong Kong. Hong Kong's history with video games is in hardware, because Hong Kong is a weird intersection between America, Japan and the U.K. Hong Kong has its own specific games consoles, which are the Hong Kong version of the Japanese one, or the Hong Kong version of the European version. So, it's actually in hardware rather than the software where Hong Kong's visual culture with video games actually is.

LARA: There's a very specific context we're working with in Hong Kong, even down to the video game console. There's no single "Asia."

DAVID: Yeah.

LARA: They're such specific histories. One of the things that we talk about at *M+* is how to define a region, like Southeast Asia. How do you respect the specificity of all the different micro or subcultures that exist within Asia, and make it part of a broader history of visual culture? When creating a new history of Asia, where do you start? Where's the entry point? *Asia Art Archive* is great because you have that depth of research, and you expect people to have the context already, and you're deepening that context.

We're currently talking about user stories or user journeys for our website and across many different experiences that we're building. Someone from my team loves Nintendo. The thing that she loves is the idea that there's one main story that anyone can have access to, and you can go from beginning to end, and that's your experience of the game. But depending on how deep you want to go, or how geeky you are, you can go deeper at various junctures. That's what, for her, makes a really successful game.

It's something that we're thinking about. How deep do we go? How do we frame it so that we can be inclusive and accessible, and create a level playing field? We can't expect all our audiences to come to our new storytelling about Asia with the same level of depth and knowledge.

KEIR: Right. The story needs to work for people who skim it, for people who dive deeply into one part and for those who swim around in it.

I want to follow up on this idea of creating a museum for Asia, based in Hong Kong. In 2012, *M+* received a major gift from Uli Sigg, the Swiss collector of a comprehensive private collection of contemporary Chinese art.

When he announced the donation, he said he chose a Hong Kong museum over one in mainland China because the collection includes work by artists suppressed by the Chinese government.

Since then, M+ has collected photos by Lui Heung Shing on China after Mao, which includes work on Tiananmen Square. How does Hong Kong's political context inform your respective digital practices?

LARA: M+ is a global museum looking out at the world from our base in Hong Kong, and the Hong Kong perspective informs everything we do, digital and otherwise. The political context here gives us civil liberties like freedom of speech under the Basic Law. We also face tensions as those liberties get tested, but it's a bit of a paradox, because the conditions that create these tensions are the same ones that make our perspective unique.

The first exhibition that we held at M+ when I joined was highlighting the M+ Sigg Collection. The curatorial approach was chronological. At first, I thought, what's the big deal about showing a chronology? That's not that revolutionary. It didn't seem that special to me.

But it was fascinating to see how people responded to a giant timeline... Looking at these works, being able to position these works in terms of their recently lived history, the context of their memories, the context of their families. To be able to place that in a timeline is something that we can offer that someone in mainland China may not be able to have access to in the same way.

We have the freedom to give access to the M+ Sigg Collection and show it even more dynamically through digital. We're building an online exhibition that lets people explore a segment of the collection through time, as well as art movements, artist groups and world events. We can tell these stories because of where we are, share them digitally and encourage people to be curious, to explore new dimensions of understanding. We'll have similar aims when building out our future M+ Collections site, which we'll be approaching as an open-ended "beta," and which the M+ Sigg Collection will be part of.

DAVID: Although AAA is based in Hong Kong, our collection remit is across Asia, or the multiple Asias. We do acknowledge some of the political things that are going on, but it doesn't stop our collecting work.

When they had the Occupy movement in Hong Kong, we had reference to that on our website, and our website disappeared in China. It reappeared maybe a month later. That's happened twice, I think.

In our performance art collection, we have material from an artist and activist and documenter called Ray Langenbach. He videoed performance artworks across Southeast Asia. This included one work called *Brother Cane*, which caused a huge tabloid panic about performance art in Singapore in the late '90s.[1] He did a performance work in reaction to a news story about

---

1 *Asia Art Archive* provides onsite access only for the video documentation of Josef Ng's performance *Brother Cane*.

some gay people being arrested and being caned as punishment. That caused a huge tabloid scare, and the Singapore government restricted performance art for 10 years. They didn't ban it outright. They simply refused to fund anything that had a performative element in it.

KEIR: Wow.

DAVID: We have that in our collection, and it is available in Singapore. Now that time has passed there are discussions about retaining copies of it at one of the Singapore art museums.

One of the censorship questions we have had has been, what happens if someone gets hold of your web server? We've had to develop a whole strategy of cloud redundancy to make sure that if we get shut down, we'll be able to continue to make material available digitally.

LARA: There were questions before we launched the *M+ Sigg Collection* show as to how it would be received. As it turns out, the connection between what we were sharing and the political implications were reassuringly separate.

Something that has also made what we're doing at *M+* easier is that the current chief executive of Hong Kong is Carrie Lam, who used to be the chairperson of the West Kowloon Cultural District. That's a huge step forwards in securing our future as an independent institution. Although we're government funded, we are not part of the government. Our remit is not the government remit, and we need to keep that distinction as clear as possible.

KEIR: David, I've seen the *AAA* collection described as a set of key areas of concern which identify thematic content priorities: art writing, complex geographies, exhibition histories, innovation through tradition, pedagogy, performance art and women in art history. I'm curious where those themes came from, and the strategy behind listing them publicly.

DAVID: When we were having our organisational rebrand, the Head of Research at the time was joking, maybe *Asia Art Archive* should rename itself "Asia, question mark, Art, question mark, Archive, question mark" because we spent so much time thinking about all these questions.

Complex geographies are a case in point. When we talk about Asia for *AAA*, we don't talk about it as a geographical thing. It's more of a concept. So, if we talk about the development of contemporary art in Asia, we have to include New York and London and Sydney, and all of these other big cities. That's started a line of inquiry about, well, how do these different art histories intersect in these other places? We've just done a partnership with the *Mellon Centre* in the U.K. called *London, Asia* that's specifically about how artists of Asian heritage have intersected in the west.

In other areas of research, we've been looking at how Hong Kong artists were doing their exhibitions in the Philippines, for example, and vice versa. How these things were intersecting. That's what's driven the complex geographies area of interest.

We did a big project in India on art writing. We tried to create a bibliography of 13 different languages across India. If you imagine the dominant art histories are an English written tradition, it challenges people to ask how art history works if there's an English one and theoretically 12 others in different languages? As a researcher, what does that mean for your research? Does it mean you have to learn 12 other languages to be able to write an art history? In a way, it's questioning some of these givens.

KEIR: I see that in your education programs. In both institutions, there's a mix of in-person and online opportunities for action, research and learning. How do you balance in-person and online opportunities? How do you measure or value them?

DAVID: Our researchers are primarily in the U.S. and in Europe. The online access to our collections is not in Hong Kong. There are a couple of reasons for that. One is that the art teaching in Hong Kong is still heavily a Western art historical narrative. It's Europe, Europe, Europe, America. But our programs and education are very Hong Kong based. We're working with schools and with universities to try and increase engagement with art – with the understanding and the knowledge around art.

For example, we're doing a partnership with one of the local universities on a Hong Kong art history semester. The students come in, pick a subject, look at key examples, like the Oil Street art space. They're coming to Art Basel.

At the school level, we're working with teachers to help them develop curricula to engage students with art in both a local context and beyond.

KEIR: Is that through retreats or the teaching labs and the residency program?[2]

DAVID: Yeah. The teaching labs and the teachers-in-residence bring teachers in to talk about how art is being taught, and collectively develop strategies to increase engagement.

Our first teacher-in-residence was a guy called Ricky Yeung. He was a practicing artist and an art teacher. He did a lot of work getting the students to express themselves and to think about other art forms beyond painting.

He showed us one of his student's works. The student lived in a typical Hong Kong flat. One day, he was looking out over his street. He went down to the parking space below and, with some white masking tape, drew out the outline of a flat. It was a really amazing work. It said so much about space and living in Hong Kong.

Our most recent teacher-in-residence is looking at some of the lower-tier schools and how they can engage.

LARA: When I was growing up in Hong Kong, art was really not valued. I got to age 16 and it was not offered. There wasn't enough demand in my school for it.

KEIR: Wow.

---

2 The *Asia Art Archive* offers a number of resources for educators including annual teaching labs, yearly teacher retreats and an educator-in-residence program. They also host an art education network and provide rich educational researches for those wanting to teach or better understand art and culture.

LARA: What would have happened if something like the *M+ Rover* had come to my school? Once a year, a mobile creative van will go to different schools. It's taken over by an artist and they run workshops. Kids observe, communicate with one another, express themselves artistically and interact with a living, breathing human being who makes their living and their life through art. I think that's so powerful.

One area that I would love for us to improve on in the digital space at M+ is mirroring the work that our learning program does. There's so much room for online amplification. Of course, we have *M+ Stories*, which is the museum's online storytelling platform. Its position is to tell stories about the museum, engage the public and learn about how our content resonates with different audiences. Even though it's a really simple platform at the moment, we're working towards expanding the platform and bringing a greater mix and depth of content under the *M+ Stories* umbrella.

In the research area, we have *M+ Matters*, which is an ongoing symposium. We bring in leading thinkers around visual culture, strategically organised to help us think through a question in our collection through internal workshops and workshops with the public.

We also have a thematic publication called *Podium*, which is a curatorially and editorially led online journal of visual culture with an open-ended structure. The first issue, which launched with the *M+ Stories* website, was on visual culture in general. We've got issues coming up on objects, geographies, time and moving into the institution.

We're asking ourselves: How do these research projects feed into these broader themes, and how can they have a life outside of the very exciting symposiums that we do? How can they feed into a bigger narrative of our publishing?

KEIR: I believe that's one of the questions that will animate digital within our sector over the next five or 10 years. We've reached a baseline for getting our collections online in a meaningful way. There's a large array of tools to quickly and beautifully display content online. What we haven't done is translate the energy, intelligence, and nuance of in-person debate or research outcomes online.

Lara, you spoke earlier about being in a place that has multiple cultures and languages intersecting. I read that your collection API is being developed in multiple languages, which is super cool. Your local language context means you need multilingual content strategies from the outset. Can you talk about your practice regarding multiculturalism and multilingualism at M+?

LARA: M+ is a bilingual institution and that needs to be reflected online. We've currently chosen bilingualism over trilingualism. We publish in English and traditional Chinese, but not simplified Chinese.

DAVID: Because *AAA* is Pan-Asia, we use loads of languages.

Our primary is English. With collection items, if it's possible, we start in the original language and then we translate. The collection parts of the website display both languages together. Other parts of the website, we use English and traditional Chinese. That is quite a political statement, because traditional Chinese is really Hong Kong and the greater Guangzhou region. The rest of China predominantly uses simplified Chinese. The website never gets a huge amount of take-up in mainland China. We're not sure whether that is because of the language.

The writing style is incredibly important as well, not just the symbols. We translated our website navigation into traditional Chinese. Our China researcher, who is from Hong Kong and was based in Shanghai, said, "That's not going to work in a mainland setting, because it will come across as almost flippant." He was saying, "That would be perfectly appropriate for Hong Kong and Taiwan, but it means something kind of different in mainland China." We had to retranslate and negotiate what the Chinese language navigation was going to be.

KEIR: Is that a specific Hong Kong or a Southeast Asian perspective? For instance, if someone working in an Australian or West Coast American museum and is seeing explosive growth in Chinese visitation, and they want to meaningfully engage with this visiting population, are there things that can be generalised from what you've learned? Or is it case-by-case depending on the makeup of the specific organisation's visitor population?

LARA: Our bilingualism is absolutely rooted in the fact that we're a Hong Kong institution.

DAVID: That is our position, too.

LARA: We're a publicly funded Hong Kong institution. Our baseline needs to be traditional Chinese. If our mission was simply amplification – maximum reach – we would need to expand into China-specific content and build that into our strategy, but that's not our immediate goal. We have to be local first.

We need to be a sustainable institution, so at this stage we cannot sustain three streams of language development. Can we write in Chinese in a way that's comprehensible to a mainland audience, and then simply convert traditional characters into simplified? Ideally this is a conversion, not a translation.

Chinese-language social media is another thing to think about. We have Instagram, Twitter and Facebook. Instagram and Twitter are bilingual. If you want to do WeChat properly, you need to be thinking about the ways that people use the internet in China, and the internet languages people use, which are completely different and always changing. We've been advised that to develop a proper WeChat presence, we need at least two people on staff, and four students in the mainland who are keeping up with internet culture and developing content and publishing all the time.

These are completely different content strategies to the way you would develop an overall museum content strategy, which is grounded in developing collections content. It's a serious resource undertaking. We also need to respond to the fact that we are going to have many mainland visitors once *M+* opens thanks to the high-speed rail link. We anticipate that *M+* and West Kowloon are going to be one of the top destinations in Hong Kong.

DAVID: We have a very similar situation. We rolled out our new website last year which is dual language. We spoke to our China researcher about it, and he said, "It's really good, but it probably won't do much business in mainland China." Modern websites reduce the amount of text and have lots of pictures, and he said, "No, in the mainland it's the other way around. Mainland Chinese users just want one big page full of text. They don't want pictures."

KEIR: Right.

DAVID: This is after two years developing a website. He also told us that ". . . everything is through WeChat. You can build the website and have it in as many languages as you want. That's not where people in Mainland China are. People are on WeChat."

The thing with WeChat in China is it has to be registered to a person's identity card.

LARA: In mainland, yeah.

DAVID: It's a bit of a legal grey area for organisations in Hong Kong to be using WeChat. There are companies that facilitate doing that, but that was something that we weren't comfortable doing. It means that we're not really there. Although we do have researchers that work in China, and they have WeChat, so they share our information.

KEIR: So, it would be like inhabiting a host person who would be your voice and conduit into WeChat as an outside entity? Even if you had the breadth of resources and the breadth of content that have had a meaningful attempt to engage with mainland Chinese audiences through WeChat, there's just fundamental structural blocks to that sort of genuine and successful engagement on that platform and with that community.

DAVID: Yeah.

LARA: That's correct. My remit is the digital program. We sit under the curatorial team. WeChat also has a commercial implication that needs to be thought through. Marketing is not strictly my job, but we need to work very closely with marketing colleagues while thinking about how this affects e-commerce. It's a big structural question.

KEIR: I think it's a good challenge, especially to Western museums, to be thinking about these questions of cultural relevance and storytelling. It's still understanding your audience and your context in a meaningful way and a segmented way, but also asking, "What are the strategic choices that I can make that don't leave my audience behind, don't leave my institution behind and meet people where they are?"

DAVID: Our detailed audience segmentation was developed so we can address each of those different audiences and get it as right as possible. For example, our academic researchers didn't particularly like doing searches if they were looking through the archive collection. They would navigate to the top level of the collection and then just systematically look through everything in the collection. So, the digital equivalent of looking through a box of papers in a traditional archive.

The other thing that they all admitted to was that they just stole pictures off the website. They just drag and dropped all of the images and the metadata off the website. One interviewee, a PhD candidate, said the lecturer told them to do it. They were told, "You don't know how long that material's actually going to be online, and even if you do, if you reference a URL and it's not a persistent URL or a DOI, then it could disappear, so just copy it all locally."

Because everything's contemporary, everything's in copyright, and copyright, for a lot of the people that we deal with, is a bit alien. . . . It's not something that they're interested in or they have any great knowledge of. If you're working in contemporary art in Vietnam or Cambodia, you could be so removed from the market that you're not doing it to make money, you're doing it to express yourself. To have us come in saying, "Oh, can you please sign this document that says you're the copyright holder and you give us a sub license to make it available online?" Their response is, "What does that mean?"

LARA: We have exactly the same thing.

DAVID: Bringing in ideas such as open licenses, which are across the museum sector in Europe, America and Australia, are not so well understood in this region, and often not implemented. Sometimes it's a long, drawn-out conversation to explain – one, the idea of rights and two, the idea of open licensing – before getting an agreement. It's just not prevalent in the region.

KEIR: There are some projects that I'd love to do that we won't even bother starting because of copyright. There's an audience and there's a business case, however, copyright has killed it before we've even started.

Looking forwards, what does cultural participation look like in Hong Kong, and what might it look like in 15–20 years?

LARA: At *M+*, because we haven't had a building up to now, we've had to be really creative about how to get out there in the community and how to inspire cultural participation. There's going to be a jump-off point where we have this physical space, but 15 or 20 years from now, that cannot be the boundary or the remit of what *M+* is. The cultural participation doesn't just happen by being inside the space. We still need to be interwoven into our communities, and we need to be out there with our content and make our way into people's lives.

We want our brand, our knowledge base, our storytelling, our ideas to be out there and to be used by people in the future, and for them to have ownership. We're moving towards an open-access release for our data in the coming year, which will be foundational to everything that we do in the future. That's my ideal projection – to still exist outside of the museum walls even after we have a building.

KEIR: Bravo.

LARA: And be relevant.

KEIR: Yeah.

DAVID: I think that for *AAA* over that period of time, I would hope it continues being a point of authority. As a node within a network of archives, museums, practitioners, artists and researchers, hopefully it's in some way a lot more connected than it is at the moment. Hopefully some of the organisations that have been working the way they've always worked, start to think forward a bit more about opening up their collections, making material available, being more comfortable with how information flows digitally. That's the kind of thing I would love in Hong Kong in 2047.

KEIR: You're talking about the full unification with China?

DAVID: Yeah, in 2047.

KEIR: 2047. We should still be around for that, climate change willing.

LARA: Yeah. The thing is, we're in this floating transitional space where there's a lot of uncertainty about the future. As we develop in the next 15–20 years, we need to be actively be a part of shaping that future, and in a way that feels hopeful, not anxious or oppressed. Those are feelings that are in the air in the Hong Kong political context right now. How can we as art spaces activate our audiences, enable them, empower them to shape the future so that they feel like it's theirs, and it's not imposed or completely out of their hands? How can creative expression in some way contribute to this, and how can we as cultural organisations enable this creative expression and cultural participation? That is important and urgent and relevant to where we sit today.

DAVID: Yeah, absolutely.

KEIR: That's a great place to stop. Thank you both.

One of the most present tensions of the digital age has been the need for ongoing negotiation between the global and the hyperlocal. There is a pervasive notion that digital technologies have democratised information and access, in ways that eradicate the challenges of physical distance. While there is ample reason to believe that this is an incomplete or incorrect understanding, given the many digital divides that shape how people are able to interact in online spaces, it is true that digital technologies have reconfigured many notions of place, time and community. This discussion between Lara and David explores what it means to acknowledge the diverse, nuanced and (at times) contradictory nature of something as multifaceted and complex as "Asia," as an idea and as a geography, whilst exploring how to be part of global museum and art making cultures, all whilst still being grounded in Hong Kong and Asia. Such questions, in some ways, reflect the tension found in collecting and memory institutions as they navigate the relationship between telling specific, local stories and connecting such stories to broader histories.

Navigating mixed contexts that range from the global to the hyperlocal also means performing regular acts of translation, especially in multilingual cities, whether that means translating content into different languages, translating living cultures into institutional contexts, or translating digital practices and technologies into a museum context.

For more on the challenges of translation, we suggest you read:

> **Conversation 2: LaToya Devezin + Barbara Makuati-Afitu**
> **Conversation 4: Sarah Brin + Adriel Luis**
> **Conversation 5: Sarah Kenderdine + Merete Sanderhoff**
> **Conversation 6: Kati Price + Loic Tallon**

For more insight into the kinds of discussions that need to be had, both internally and with external stakeholders and communities, when building a new institution or implementing significant change, we suggest you read:

> **Conversation 7: Shelley Bernstein + Seb Chan**
> **Conversation 9: Brad Dunn + Daryl Karp**
> **Conversation 11: Tony Butler + Lori Fogarty**
> **Conversation 12: Daniel Glaser + Takashi Kudo**

Reference

Asia Art Archive. 2017. "About the Asia Art Archive." aaa.org.hk. https://web.archive.org/web/20171219034636/**https://aaa.org.hk/en/about/about-asia-art-archive**

CONVERSATION 4

Sarah Brin + Adriel Luis

"If your primary goal is simply relevance itself, then there's not really a foundation upon which that relevance is built."

– *Adriel Luis*

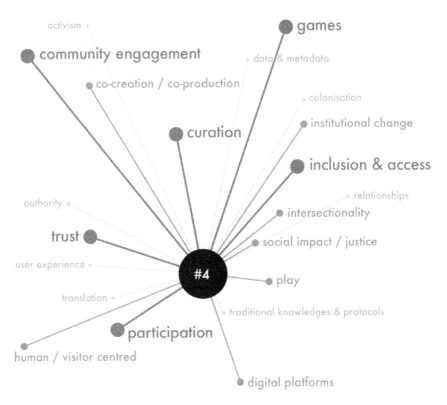

FIGURE 4.1

Both Sarah Brin (U.S./Denmark) and Adriel Luis (U.S.) joined the museum world via circuitous paths and work in progressive artist, visitor and community-centred ways. Sarah's experience working with gaming and technology-focused pop-ups and ephemeral curatorial projects in the museum and commercial sectors led to her participation in the Horizon 2020-funded GIFT project that explores hybrid forms of virtual museum experiences as a researcher and development manager at *IT-Universitetet i København* in Denmark. In 2013 Sarah co-curated *Ahhhcade*, a relaxing game pop-up exhibition at *SFMOMA* and in 2017 she helped coordinate the *PlaySFMOMA Mixed Reality Pop Up Arcade* which Keir's department also contributed to.[1] Soon after this conversation was recorded, Sarah joined *MEOW WOLF* in New Mexico, U.S.

Adriel, a self-taught musician, poet, curator, coder and visual artist, developed the innovative *Culture Lab* series with his colleagues at the *Smithsonian Asian Pacific American Center*, a museum without a building. *Culture Labs* are "fleeting, site-specific happenings that recognise art and culture as vehicles that can bring artists, scholars, curators, and the public together in creative and ambitious ways" and have been held in Washington, New York, Hawai'i and New Orleans, following the tenants expressed in the *Culture Lab Manifesto*.[2]

We paired Sarah and Adriel to discuss agency, expertise, play, trust and institutional change, both inside the traditional museum context and beyond, contrasting American, European and Asian perspectives where possible.

This conversation was recorded on January 4, 2018. At the time, Sarah Brin was Curator and PhD Fellow at the *IT-Universitetet Kobenhaven* in Copenhagen, Denmark and Adriel Luis was Curator of Digital and Emerging Practice at the *Smithsonian Asian Pacific American Center* in Washington, DC, U.S.

KEIR: Sarah, could you describe the GIFT research project – its methodologies and goals?

SARAH: GIFT has a bunch of different components, many of which are aimed at connecting visitors to exhibition content through playful or interactive experiences. We work with artists and designers to develop prototypes and have a team looking into the theoretical background supporting these ideas.

I'm working with a consortium of different types of international museums in an 18-month design research process. During our time together, we reflect upon and identify some of their key goals, their target audiences and the assets within their organisations. After we do that, we do some

---

1 Both events were produced by *SFMOMA*'s Interpretive Media department as part of the *PlaySF-MOMA* initiative that encourages the development of avant-garde and experimental games.
2 The *Culture Lab Manifesto* was first published in 2017 in *Poetry Magazine* by the staff of the *Smithsonian Asian Pacific American Center* (Luis et al. 2017).

experiments that are intended to serve some of those goals while simultaneously developing capacity within their organisations. While we're designing these experiments, it's really important for us to emphasise specificity. These small-scale experiences are designed to get buy-in from other stakeholders within the institution.

It can be tempting to want to design an exhibition project or experience that can somehow magically serve and please every audience. But it's important to be more targeted than that. It's crucial to have a clear idea of who you're trying to serve when you develop a new initiative, especially if you're doing something that could be costly, like rolling out a new technology in an exhibition space. That's a lot of the rationale behind creating user stories, which is a common practice in the fields of interaction and web design.

But non-experts, people who might not feel at home in the museum, are one particular audience segment that I'm interested in. Most of the programs I design (in my own curatorial work) are specifically targeted towards them.

KEIR: Something we talk about a lot at *SFMOMA* is trying to move from a language of authority to a language of expertise, and broadening the number and the diversity of the people who are allowed to have expertise.

ADRIEL: Making meaningful connections, particularly to people who are not conventionally considered experts, is very much at the centre of what we've been doing at the *Asian Pacific American Center*.

KEIR: Tell us more.

ADRIEL: I joined in 2013 under this title of Curator of Digital and Emerging Media, a position that was brand-new to the *Smithsonian*. From the job description, it was clear that they were looking for someone to address a perceived divide between physical and digital museum experiences. We recognise that museums grapple relevance and knowledge. Public concepts of science, art and culture are changing rapidly, often unexpectedly. How can a museum respond to this? At the *APA Center*, our response has been the Culture Lab.

A Culture Lab is a community-centred alternative to museum exhibitions – where from the ground-up, everything is developed as a practice in community organising. The Culture Lab is different from a conventional exhibition, because instead of curators telling you the answers, curators ask questions. Visitors to Culture Labs encounter conversations that have not quite reached an answer or consensus yet. The curators, artists and cultural practitioners create modules in the form of art, or workshops, or other formats, that address big topics such as race and equity, or the environment or media representation. Each Culture Lab takes on a theme that can encompass a plethora of pressing issues, and lets the artists and creators run wild. As of 2018, we've done three Culture Labs. We did two in 2016 and one in 2017, respectively, in DC, New York and Honolulu.

KEIR: Looking at these projects, I see this shared through line of non-traditional or experimental approaches to conceiving and creating an exhibition outside the "museum exhibitionary industrial complex." From my understanding, you both value this inter- or trans-, or even anti-disciplinarity. Sarah, could you talk more about those ideas?

SARAH: A lot of the work that I do involves functioning as an interlocutor; someone who, like Adriel, facilitates a conversation. Being an interlocutor doesn't necessarily mean being an expert, but a lot of the work involves scaffolding a series of questions, and highlighting overlaps and dissonances between artists' work and audiences' perspectives and experiences. That can definitely resemble a translation process, sometimes.

When I was in the private sector at Autodesk, I did a lot of translating the creative research that artists wanted to do, explaining how it was a good economic investment for a software company. Also, conversely, taking the internal language of the software community and the capabilities of particular piece of software or machine, and communicating that to non-expert audiences, making those processes that are hidden by code or by the surface of a digital fabrication machine a little bit more legible.

KEIR: Adriel, I was rereading the Culture Lab Manifesto recently – and I love that you published that in a poetry journal, by the way. You reference Kimberlé Crenshaw's theory of intersectionality in the manifesto. I'd like to better understand the value of this thinking. How do you apply these goals, these elements of the manifesto? How do you operationalise those aspirations?

ADRIEL: The Culture Lab Manifesto is our first step towards presenting the Culture Lab as an open-source model that others can adopt, whether they're a big institutional museum, or a small non-profit or gallery. My co-curator, Lawrence-Minh Bùi Davis, and I both come from literary backgrounds. He's a writer and editor; I come from spoken word poetry. We knew that if we published in a museum journal, only museum people would read it. For us, publishing the Manifesto in a widely read platform like *Poetry Magazine* means that we are curating for our communities and not just for our colleagues. In May 2018, our education specialist Andrea Kim Neighbors led the publication of a playbook, which more deeply explores the beliefs outlined in the Manifesto.[3]

The Manifesto came to us when we realised that sharing the Culture Lab model shouldn't begin with technical strategies like how we go about paying artists, finding our venues or certain aesthetic things that we abide by. Instead, we wanted to make very clear our values that lead to such methodologies. If someone wanted to put together a Culture Lab, they wouldn't be

---

3 The Culture Lab Playbook was published online in 2018 and is available to download from https://smithsonianapa.org/culturelab/.

under an illusion that you could replicate one through aesthetic. You could book all the same artists and have them do the same work, but if the ways that the artists are treated, the way the topics were conceived or the reason the city was chosen have not been considered through the lens of ultimately uplifting communities, it wouldn't be a Culture Lab.

Early in my career at the *Smithsonian*, I learned that there is no shortage of rules, but it was very difficult to find values. When it comes to curating an exhibition, I can find reams of stuff around fire code and how to position our logo. But it's more difficult to locate our community values – how do we treat each other? How do curators treat artists? How do curators treat admin? When someone experiences one of our projects, how do we want them to grow? Maybe the closest that we might get to an idea of values is the spirit of knowledge and knowledge seeking at the *Smithsonian*.

I learned that curating led by community principles is counterintuitive to museums. Museums are historically instruments of colonisation, particularly in non-Western regions. The first museums in Asia were in India, founded by the British, and in Indonesia, founded by the Dutch. These museums were founded as tools to present narratives to people of colour via their colonisers. How can we move forward, assuming that nowadays museums present themselves as public spaces . . . and under the mis-assumption that everybody trusts museums? These issues that people have with museums haven't been solved. These questions led to the theme of intersectionality for our first Culture Lab, *CrossLines*, and continues to be explored and debated through Culture Labs by recognising that institutional knowledge should be informed by the increasingly complex ways that culture and society shapes our collective understanding of humanity.

KEIR: What you're describing sounds to me more like the community organising principles you'd find in grassroots or local politics rather than in traditional curatorial practice.

How different is this from your experience at Autodesk, Sarah?

SARAH: When you're working with contemporary art and with technology, timeliness is essential. The thing about working in the private sector is that it can move so much faster than a traditional museum. Because there's that imperative to reach audiences and to be timely and responsive to ultimately sell things, you have that luxury of moving faster. Big decisions usually go through just a couple rounds of approval. If you're working at a museum of any size, you have to plan your exhibition schedule at least 18 months ahead, and more realistically, you're looking at three or four years. Of course, lots of museums are smaller and can work more swiftly than that.

A museum will not necessarily make it a habit to consistently acquire newer artwork made with technology, partially because they're hard to collect and maintain. . . . Well, the commercial models are different. I find inspiration from artists who are finding new ways to survive and support themselves. You have artists who have day jobs making multimedia spectacles for Doritos, or whatever other brand, and in their free time, they can support their art practice using the same or similar art-making tools they use to produce client work. They're not trying to work through a commercial gallery system. That kind of adaption is inspiring to me.

KEIR: I think a lot about the default temporal modes that museums lumber into, thinking about objects on the 10-year, or 100-year time scale. There's exhibition time, where planning is 18 months to five years. Then you think about digital practice being at an incredibly high tempo. Because museums don't traditionally operate at that tempo, they often just apply either exhibition time or object time to the production of digital work. It's an obvious friction.

ADRIEL: I just read a book called *The One Straw Revolution* by Masanobu Fukuoka. He was a farmer who, in the 60s and 70s, inspired permaculture movements. I'm connecting a lot of what Fukuoka is saying to something that I heard from Stewart Brand, who wrote *The Whole Earth Catalog*. Both talked about change based on speed and scale. How things that change really quickly, like fashion, often have fleeting consequences, and things that change rather slowly, like institutions and the climate, are of major consequence.

As someone who grew up in Silicon Valley as an 80s baby, that's definitely been one of the challenges I've had for the last few years working at the *Smithsonian*. What does it mean to be working at an institution? What does it mean to accept that some things, many things, move slowly? What does it mean to be sitting in the office, staring out the window, feeling like the rest of the world is just moving a lot faster? What Brand said, and what's embedded in this idea of permaculture, letting nature run its course, is this idea that there's definitely value in being associated with things that have outlived any human being on Earth.

There's a certain kind of knowledge, a certain kind of experience, that's difficult to embrace in the moment because everything else seems to be moving really quickly, right?

KEIR: Right.

ADRIEL: In the museum world, we've come to accept that slowness and responsiveness are mutually exclusive. I think it's just as harmful to be completely inert and not able to respond to something, as it is to scramble to respond to something quickly – to be the first one on this new platform, or the first ones to break this news.

We've seen in other fields, like in the press, what happens when you just want to be the first, or just want to respond quickly to be relevant. Sometimes you end up following the wrong rabbit hole in the worst of situations, like disseminating terrible or inaccurate knowledge. I think that's something really important for museums to grapple with. I'm hearing from a lot of people who are interested in rapid movement as the solution for relevance, and I think that we need to step back and think about where rapid movement falls into the strategy of being responsive. It does, but it's not a silver bullet. What does relevance in the moment mean for the long-term? If your primary goal is simply relevance itself, then there's not really a foundation upon which that relevance is built.

KEIR: That's super interesting. You mentioned trust earlier, but you're also talking about how trust can be eroded through a set of activities. I think that's the case in politics and the media in particular. Public trust is something museums, libraries, and archives have and what politicians and media organisations often lack. If there are only a few places left in our communities, such as museums and libraries, where there is consistent trust from the public, does that change our role?

SARAH: Most definitely. It's more vital than ever, as these interlocutors associated with public institutions that people still trust. It's our obligation to be the translators, and to create openings, approachability and understanding within these seemingly impermeable institutions. It is not the time to be didactic. We need to think more about the strategies responding dialogically to the communities we serve and shape. We need to lean on that public trust to gather more information about how to craft exhibitions and museum programs; how to craft these narratives about what culture is, and what it has been and what it can become.

KEIR: Adriel, as a free, publicly funded institution, is it the *Smithsonian*'s place to fill this information gap or this knowledge confidence gap?

ADRIEL: Yes and no. I'm still thinking about this idea of relevance and of trust. In 2016, right before we opened *Crosslines*, I wrote about the way that I went into curating with trust in the artists and asking artists to trust me in a way that they hadn't before been asked or inspired to trust curators. Before working in museums, I was a working artist. I was also working in non-profit organisations as a consultant, and sometimes as an employee. When you're working with grassroots organisations, or community organisations, you never go to a meeting where people are asking, "How do we become relevant to the community?" because you're a part of the community.

When you go home for the holidays, you don't ask, "Well, I've been away from my family for so long, what do I say or do to be relevant to it?" Because you're a part of that family. Community organisations evolve, they change and sometimes they lose trust, but as long as they're doing programs with the community at the centre, that never becomes so much of an issue.

I think museums are going through this interesting moment right now because they're realising that even though they've been public spaces for quite a long time, it's a relatively new idea to meet the community at an equal level, to serve certain members of the greater public that in the past they never were asked to serve. At the *Smithsonian*, we have a couple of really interesting vantage points to come from. Because we are so multidisciplinary, it's not completely unexpected for us to carve new formats.

The exploration of new concepts and approaches for knowledge-sharing has been expected of us ever since we've been around.

That said, growing up in California, I thought the *Smithsonian* was a single museum that I might never visit. I try to keep that in mind when I think about the people that I'm outreaching to. Maybe some are sceptical of institutions in general, or museums, or anything that's associated with the government. Others might come at it from a whole other side, assuming that the *Smithsonian* has hired all the world's experts, and so anything that they see at the museum is going to be absolute truth.

KEIR: Hmmm.

ADRIEL: When I hear people in the museum world say, "How do we get the public to trust museums," I wonder what it is that they exactly mean by that. Do they want people to come into the exhibition halls and never question what they read in the labels? That's not the kind of trust that I think is most important. I think what's important is creating an atmosphere where people trust us to make it conducive for them to trust each other. How do we create peer-to-peer learning in museums?

I definitely agree with Sarah that we're now past the point of the public trusting museums without question. I don't think it's necessarily productive for us to create or re-create a situation where people give us the trust that, say, they gave the news when TV was first introduced. We're living in a very different time right now. Trust is going down, not just for institutions, but in general.

The last bastion of hope when it comes to trust isn't in museums or institutions, it's in each other. How do we cultivate museums as environments where people can have the kinds of conversations that they might not have anywhere else?

KEIR: I think of museum trust as more akin to the trust that you have in a good friend, in that you can trust them to be consistent and authentic, but you can't trust them to be right all the time. It should be a passionate, nerdy, generous expertise.

Sarah, as an American now working in Europe, on a project that's pan-European in its nature, do you notice a difference in how audiences you're engaging with, and practitioners you're working with, conceive of these questions? Does the publicly funded model of museums change how people perceive them? Do they trust them differently; do they engage them differently? What are you seeing in-between those geographies?

SARAH: I think there's a lot of dismay and frankly, grief, [around Brexit] looking at the way democratic processes have been supposed to serve a country, and immense amount of fear about not knowing what will happen next . . . If cultural institutions that have historically been funded in some capacity by the government will continue to exist. If they do continue to exist, in what ways?

I am heartened by a sense of kinship . . . Museums are working to create scalable tactics that can be implemented at different types of institutions, working at different operating budgets and at different sizes. It's almost like everyone is working in crisis mode and figuring out, "How can we plan for a catastrophic future, and how can we support our colleagues doing similar work?"

Of course, within the EU, cultures and cultural heritage institutions vary wildly from country to country. I don't see anything even vaguely resembling a pan-European identity.

KEIR: Adriel, you frequently travel to Asia and the Pacific, a region home to half the world's population, and nations, major faiths and languages. Are there things you're seeing in Asia, in the structures that are emerging, that the West should be aware of? What is the inspiration you find there?

ADRIEL: I'm really excited to talk about that. When I first came to the *APA Center*, we were just looking at the experiences of people of Asian and Pacific descent who are located within the territories of the United States. But Asian Pacific America isn't just a domestic American experience, it also the history of colonisation, migration and diaspora. What's our future in a more globalised and digitised world, where communities and identities aren't necessarily confined to geographical borders and citizenship?

Before I joined the *Smithsonian*, I was living in Beijing and travelled all throughout Asia. I started a tradition of going to museums, and discovered that local art is my favourite way to learn about a new place. I grew up with a very American assumption that China has a completely neutered art scene. With all the censorship, how could anything provocative, or expressive, or cutting-edge ever emerge? On the contrary, I found really compelling, nuanced, subversive work that would never have been thought of in the U.S. because of the different guidelines for expression. Within constraints of expression, some amazing creativity flourishes – and this manifests in different ways throughout Asia.

A curator named Zoe Butt has written about how Vietnam's gallery scene, particularly in Ho Chi Minh City, has created a permanent model of temporality. At any point, an exhibition, artist talk or program could trigger the government to shut down the gallery. So, the galleries have had to learn how to recognise that any day, they might have to move. They've had to figure out how to communicate to the public that they are not defined by their building.

In Cambodia during the Khmer Rouge, pretty much all of the artists, artisans and cultural practitioners were killed or forced out. For an art scene to emerge from that is like a rose out of concrete. Over the past few years, galleries have opened in apartment buildings, in underground spaces. Even with all the aesthetics often stripped down, even when labels need to be written on Scotch tape, and paintings are curling in the elements in ways that would appall U.S. museum practitioners, the stories of the communities still thrive.

And in Thailand, if you're creating new media in a place where your electricity cuts out every day, you don't have access to really expensive and innately impressive gadgets, and you really have to leverage your creativity.

I think it's important for the Western world to observe what's happening in Asia at this particular time, and to recognise how incredible it is.

KEIR: Wow, that's interesting.
ADRIEL: Yeah. Yeah.
KEIR: Sarah, how are you finding museums can bring play into their visitor experience strategies? Is there some fear within some organisations that play reduces the public trust in institutional seriousness and academic rigor? What are the possibilities there, and where do you address that internal bias and internal fear when you start looking at organisations?
SARAH: One of the missions of my research is empowering more people to have opinions, and sometimes that means unpacking negative, critical or sceptical perspectives. That's why I like working with popular media, especially games. I am very familiar with institutions' fear of games and playfulness; there are a lot of assumptions that these things are just for kids, or that they're low culture and so on.

I love using prestige transfer as a way to deal with that; by applying the same kind of academic prestige that museums have leaned on historically, pulling in historical examples of artists who've been using games, as well as looking at more established academic texts, either in the field of game studies or in psychology, et cetera, et cetera, to support contemporary work.

I'm working with museums that operate in languages other than English. When you talk about a term like playfulness or games, that can mean different things depending on the cultural context. A lot of that work is figuring out how these concepts translate and then trying to figure out what that exact anxiety is, and addressing it with examples or perspectives. If you look at playfulness as a perspective, as something that might mean something child-like or recreational, you can also look at it as trying on different identities. You can look at it as an experiment. You can look at it as thinking creativity and contextually at the same time. I think that an institution will usually come around to it, but they might not always find it interesting for the same reasons that I do.

Sometimes, I can present all that information to an institution and emphasise the approachable qualities of games; the fact that they can be a great access point for talking about art. And the museum will move forward with a games program, but maybe not because they're interested in framing some games as artworks, but because games can be good marketing initiatives and they get more people through the door. That doesn't matter.

In my experience, you have a bunch of different stakeholders in museums, and they all have different visions, objectives and perspectives. Part of what I do is creating answers that meet those perspectives simultaneously, while ultimately making sure that I serve the values that I think are crucial.

KEIR: Stakeholder management and clear narratives that groups, departments or individual people buy into and support is the dark art of museum innovation success. The retail politics of making new things in traditional institutions is a lot about the language you use for different groups, so that they realise the value according to the metrics they define as success.

The projects that you've both worked on require genuine institutional change to be successful. In my experience, changing museums is difficult and slow. Adriel, how have you found bringing these sorts of mindsets and these practices into institutions? What are the barriers you've experienced, what approaches you've tried that have worked or haven't worked?

ADRIEL: The short answer is just to learn to be a bureaucracy ninja.

KEIR: I like that.

ADRIEL: I've been in museums for about five years, and I'm still very much a neophyte. I came to museums because I recognise how important they are in a larger equation for community progress.

That said, I come across a lot of people who have been working in the field for a long time, who are motivated to do something because it'll be "good for museums." When the primary objective is serving a museum, as opposed to a community, there's no foundation. Over the last few years, museum conferences everywhere have gravitated towards the theme of social change. But I find it very problematic if the incentive for talking about social change is to bring more people into the door of museums. If you were to present that theory of change in an activist circle, you'd probably be escorted out immediately.

We have to think about some kind of exchange. What is the true thing that museums are giving back? I go to these museum conferences where social change is discussed and I seldom find activists, community organisers, or people who are experts in the subject matter at hand. Instead, I find well-meaning museum professionals who can talk a lot about why their museum isn't racist, but I'm interested in seeing something go further in that.

I recognise that museums can play an essential role in social progress, but that they're not the only answer. I don't need museums to change completely. We just need to find the openings where we can leverage the power of museums to contribute to the change. For example, museums can play a role in increasing empathy in communities – but no one's asking museums to be the primary instigator of that. I would waste a lot of energy if I thought I needed to change all museums in order to contribute to increasing empathy in communities.

At the *APA Center*, I have a great team that's open and excited about instigating this kind of change. I'm really grateful that the *Smithsonian* and its museums have embraced Culture Labs. I think this kind of influence on museums is amazing, but that's not my priority. That said, I do try to open up pathways in my curatorial process, so the next time someone wants to do something towards community building in museums, they'll have found that some of the obstacles have been removed.

KEIR: There is a famous Facebook mantra of, "Move fast and break things." It's specifically museums' role to not move fast and break things, it's first "do no harm." I often think about an idea I heard from Laura Raicovich, the Director of the *Queens Museum*, at a seminar *Does Art Have Users?* in 2017 where she talked at length about making your museum's values operational.

Museum mission statements are often hard to articulate. You can look at that as an identity problem. But it also leaves an enormous amount of room to define how the work that we do, the people we hire, the way we pay, the way we treat our audiences, and the generosity with which we present our expertise can represent our values.

I feel projects like GIFT and the Culture Lab, are expressions of the activism you can do in the space that's created by a lack of distinct mission and vision, or in places with strong values.

SARAH: I do a lot of research on institutional change these days, and I find that people are always expecting to have the conversation about what an institutional mission is. People get quite surprised when I ask what success would look like to them as an individual. Institutional change, as we all know, can be a battle. Reflection is a huge part of that process, and that's part of why we ask these questions about, "What would success look like to you?"

Toggling back to what success looks like for an institution at large, I'm grateful that I get to lean on institutional prestige because a lot of our partners for GIFT are very happy about the fact that they're working on an EU research project. While normally this type of work might have to be a little bit more fringe-y, or done in secret, or done with a smaller budget, they get to come to their institutions and say, "We're working on this prestigious project."

Part of what we try to do, in order to embed institutional change, is to develop qualitative metrics for these projects and experiments that can be used to demonstrate to people who might not immediately understand the cultural value of their content why these might be valuable changes to implement.

KEIR: I was given advice early on in my career in museums by one of my mentors: "If you want to do something that is fringy, and ambitious, and audience focused in its origin, then get a grant for it, so it's hard to cut out of the budget."

SARAH: Yep.

KEIR: "And make sure you write in evaluation into the grant, so that not only are you able to do the work and get a rigorous evaluation at the end." Then once you've done the evaluation, you've got something you can use to prove that the thing worked and that you can do more. That's advice I've tried to lean on since.

Adriel, looking at the recently drafted strategic plan for the *Smithsonian*, there's a series of goals about strengthening and democratising the work of the institution. There's one in particular I want to call out, which is to reach one billion people a year through digital. I would love you to put that comically ambitious goal in context.

ADRIEL: There's a lot in this idea of reaching a billion people that I'm not sure I've figured out quite yet. Once again recognising that museums are historically tools for colonisation and imperial expansion, I'm always suspicious when museums seek to expand.

If there is any kind of institutional change that I would hope to influence in museums, it's to question that colonial approach of expansion, of reaching everything that's under the sun. It gets especially dangerous when museums rely on platforms online and offline that may or may not be worthy of public trust. It's important for museums to do their due diligence and ensure that they're not continuing a tradition of expansion at the expense and detriment of the public.

KEIR: Yeah.

ADRIEL: If you were to build a museum, there would probably be questions like, "Is this on a fault line? Is it prone to a mudslide? Is it near floods?" Right? When we build exhibitions, we have the paperwork to show the egress and safety guidelines. It's basic practice to make sure that the museum isn't constructed to put its community in harm's way.

But when we build websites or online exhibitions, or we run our social media, we don't consider that digital space with the same vigilance. When Snapchat first came out, I don't know how many museums read the terms and services before deciding to join and try to get their visitors onto it, too. It's not about Snapchat specifically, it's not about Facebook, it's not about Uber, it's not about any individual company and whether or not I trust that company. It's about where museums are directing people to in digital space, and whether museums understand the environment within which they're leading their communities.

A big source of this negligence is the desire for museums to be relevant. We believe that since everybody else is doing Instagram stories, we should do Instagram stories, too. So, if the goal is to get a million follows or a billion people, then online outreach is a very effective and plausible way to do it. There are plenty of entities out there that already have their billion engagements.

But I know that the Secretary [of the *Smithsonian*] David Skorton[4] is not about just getting the hits. I have faith because of what's written in the rest of the strategic plan that it's actually about transformational change.

---

4 Dr. David J. Skorton served as 13th Secretary of the Smithsonian, serving from July 2015 to June 2019.

While museums are trying to adjust, the way that people are consuming media is constantly changing. The public expects museums to take the higher ground and to see things from a more zoomed-out lens. If what we do at the *Smithsonian* can make some kind of cultural shift in the ways that we all look at the world, or treat each other, then I do feel like a billion people reached can be a real thing. We might not have the analytics to prove it, or there might not be a sunburst logo at every single moment where that shift happens, but I think that's how that success would happen.

KEIR: What technologies and practices do you foresee having a genuine effect on the museum sector 15 years out from now?

SARAH: I think we're going to continue to see a trend towards interdisciplinarity, connecting stories over multiple channels and multiple media. The more permeability we can work towards, the better. I think we're going to see more of an understanding of what museum APIs[5] can do and what kind of weird experiments will happen when we have more public data in our museum collections.

I do believe that we will see more work in our museums that's produced with digital fabrication tools. And I think that we will move towards more specialised hardware for interactive experiences, with a lot more VR and AR experiences. I have mixed feelings about those. Honestly, what I hope happens is that there's less emphasis on the technology itself. I see a lot of people and organisations feeling this anxiety about "How can we be relevant?" and trying to deal with that by adopting a particular type of technology. That's actually a huge mistake, because that technology is going to be outdated soon, and technology, more often than not, amplifies preexisting organisational problems.

What I do hope we'll have is a more nuanced and sophisticated relationship to technology. Hopefully, an easier or more streamlined way to share information, both internally within institutions, as well as between institutions and the broader public. My hope and my belief is that we're not going to see any one, or two, or three types of technology being more dominant in these museum spaces, but we will see less apprehension around the integration of new technologies and less fetishisation of new technologies. More of a critical approach to, "How can we use this to serve our publics, how is this pushing a conversation forward, what kind of groups, what kind of feedback loops are we generating, both internally and externally?"

KEIR: I hear that. Adriel, what do we look like in 15 years?

ADRIEL: Museums are going to continue to recognise that they need to exist beyond just their own walls, but I don't think it's going to necessarily mean

---

5 API stands for Application Programming Interface and is the most common way for computer systems to share structured data with each other. A museum collection API, for example, might allow a display system (a website or app) to find all objects created in a certain year, made using a certain material or authored by a specific creator.

that everything's going to be uploaded online. The way that we get our music, shows and news has drastically shifted in ways that opens up access, and we need not be fearful of experiencing and adapting to similar changes with museums. Since museums tend to move more slowly than other fields, we get to see the changes in other arenas before they hit museums. A lot of the changes I'm seeing now in museums are similar to what record labels were dealing with 10 years ago. The internet didn't kill music, and the record labels that survive are those who are able to adapt to the new ways that people want their music offered. Music is still largely digital, but at the same time, cassettes and vinyl are making a comeback. The traditional museum can continue to thrive while digital renditions simultaneously and mutually evolve. Hopefully, in 15 years, we'll have a generation of people who recognise museums by the experiences they provide, not just what's in their buildings.

KEIR: I want to work in that world where institutions are differentiated and they're authentic – where they're providing something that helps transform people's perceptions of themselves, of reality, of community, of possibilities, and with those that are less interested in destinations, thresholds and, as you say, dusty hallways.

Thanks Adriel, and thanks Sarah.

One of the most critical questions to emerge from this conversation is how digital practitioners in museums can act to ensure that their digital activities do not put individuals and communities at risk. This question has been complicated by the increasing dominance of surveillance capitalism, wherein big and small data can be collected, mined and algorithmically analysed for insights.[6] It is now incumbent upon digital practitioners in museums to be mindful of issues related to data collection, storage, and governance, including consent and agency, data privacy and algorithmic discrimination. It is also important to be clear about what we ask of our audiences when we partner with technology companies whose own practices may be driven by different values and ethical concerns. We must therefore ask, how do our institutional practices contribute to oppressive online environments and spaces, and who is most at risk from the decisions we make around our digital practice?

For more on vulnerability, risk and digital practices in museums, we suggest you read:

Conversation 1: Seph Rodney + Robert J. Stein
Conversation 2: LaToya Devezin + Barbara Makuati-Afitu
Conversation 7: Shelley Bernstein + Seb Chan

---

6 See the opening chapter of this book for more discussion about this topic.

For more on community-centered practice, we suggest you read:

> **Conversation 2: LaToya Devezin + Barbara Makuati-Afitu**
> **Conversation 8: Kate Livingston + Andrew McIntyre**
> **Conversation 11: Tony Butler + Lori Fogarty**

Reference

Luis, Adriel, Lawrence-Minh Bùi Davis, Nafisa Isa, Kālewa Correa, Jeanny Kim, Hana Maruyama, Clara Kim, Nathan Kawanishi, Emmanuel Mones, Desun Oka, Carlo Tuason, Lisa Sasaki, Andrea Kim Neighbors, Deloris Perry, and Emily Alvey. 2017. "Culture Lab Manifesto." Poetry Magazine. July 2017.

CONVERSATION 5

Sarah Kenderdine + Merete Sanderhoff

"Any new technology is just a mirror of humankind . . ."

– *Merete Sanderhoff*

FIGURE 5.1

Before joining the Laboratory for Experimental Museology at *EPFL*, Sarah Kenderdine (Australia/Switzerland) was the founding director of the Expanded Perception and Interaction Centre (EPICentre) at the *University of New South Wales* in Australia, and previously the director of the Laboratory for Innovation in Galleries, Libraries, Archives and Museums (iGLAM) at *City University of Hong Kong*. Sarah has long pioneered new research focused on artistic visualisation techniques for museums and other public spaces.

Merete Sanderhoff (Denmark) is responsible for *SMK*'s open-access policy, working to encourage reuse of the museum's digitised collections for research, learning, knowledge sharing and creativity. Beyond her work at *SMK*, Merete convenes the Sharing is Caring conference series, and she writes and speaks internationally about collaboration, sharing, access and data in the cultural heritage sector. She received the Danish Open Data Award in 2018 for her work in this area.

Both Sarah and Merete are based in Europe working on and within museums, as well as playing leadership roles in pan-institutional organisations such as Europeana and the Australasian Association for Digital Humanities. Their conversation includes concepts such as openness, data sovereignty, ephemeral culture, participation, empowerment and risk – and the technologies that are used to facilitate or express these concepts such as data translation, APIs, virtualisation, visualisation, simulation and immersion.

This conversation was recorded on June 21, 2018. At the time, Sarah Kenderdine was Professor of Digital Museology, *École Polytechnique Fédérale de Lausanne* **(EPFL), Switzerland and Merete Sanderhoff was Curator of Digital Museum Practice,** *National Gallery of Denmark* **(SMK), Denmark.**

KEIR: Merete, could you describe what a Curator of Digital Museum Practice is?
MERETE: I dubbed the job title myself. My background is as an art historian. I wrote a dissertation on the institutionalisation of the avant garde in the 20th century. I was focusing on critique of the canon and how canonisation is developed over time and becomes a power factor on the art scene and in the way we conceive of art history. This puts a lot of the items in our collections into the periphery, because they're never on display and hardly ever mentioned. Digitisation can bring all of this material to the surface. With the internet and search mechanisms, for the first time in history the general public has the possibility to see all of the cultural heritage that's accumulated in institutions. This is a huge breakthrough in how museums can fulfill our mission as learning institutions.
KEIR: Merete, how would you differentiate yourself from a curator of born digital objects or media art?

MERETE: As a curator, you are responsible for a certain field of your museum's practice. Some curators are responsible for 18th-century paintings or Danish graphic art. I'm responsible for Digital Museum Practice. It has nothing to do, *per se*, with digital-born works in the collection or on display. It has to do with the practices that we can use to make optimal use of our digitised collections.

My full title is Curator and Senior Advisor in Digital Museum Practice. The advisor role is one that I got after specialising over a period of time in this field. Instead of researching art historical material, I research digital museum practice, and then use that knowledge to advise the museum directors, my *SMK* colleagues and colleagues within the field, both in Denmark and abroad. *SMK* is the national gallery, and we have an obligation to help develop museum practice in Denmark on many different levels, including digital.

In 2016–18, I've been the Chair of the Europeana Association Network, and currently serve on the EU Commission's Digital Cultural Heritage & Europeana Expert Sub-Group. Europeana is the platform for all of Europe's digitised cultural heritage. It has a democratic governance model where members of the network can be elected and join in its governance. That results in a community of cultural professionals from all over Europe who contribute to defining the overall strategy of European digital heritage development. My engagement in Europeana is a really important strand of my work.

KEIR: Sarah, how do you define digital museology?

SARAH: I'm interested in the nomenclature of "Digital Museology" or "Digital Museum Practice." When I came here, I was quite resistant to taking the title Professor of Digital Museology, and said, "surely it's post-digital that we are dealing with." The directors said, "Oh no, you couldn't possibly call it post-digital because the scientists here would have no idea what you're talking about." That was an interesting problem, because I feel that we live in this post-digital world where many digital processes are in operation, which might signify that we no longer need to specify digital humanities, digital museology, et cetera.

MERETE: I think it's a transitional thing.

SARAH: Yes.

MERETE: For the potential of digital technologies in museums to really take off, digital needs to be completely integrated into a museum's overall strategy, into the mindset of what we do. To paraphrase Jasper Visser – it's about attitude, not technology.

SARAH: Yes. Philosophy, not technology.

MERETE: Digital has brought about an openness, a many-faceted dialogue – a flexible, dynamic way of working. These are features of digital. It's not the technology itself. That's just an enabler.

SARAH: Yes.

With my current role, I've moved outside of a specific museum context, but I've worked in a hybrid situation across multiple museums. I haven't been an academic for very long, but I moved into my role at *EPFL* because I found I could better contribute with more firepower, with more money, on certain ideas that I have about how we can enact digital culture in museological space. It's changed how I can access resources such as high-end grants, do multi-platform projects, and engage with leading research-based computer scientists.

A specific example of this kind of hybrid role between the university and the museum is a project like DomeLab. DomeLab is the highest-resolution touring fulldome system[1] in the world. I was able to write a grant with 11 organisations, including the network organisations, other universities and three major museums in Australia for a $600,000 system. We share the system, and bootstrapped all of the technologies needed to work in these spaces. Different partners are contributing software and the dome is touring around the world.

We just did an installation of DomeLab with a project for *The National Museum of Australia, Songlines: Tracking the Seven Sisters*. This was an installation embedded into an aboriginal painting show that tells an Indigenous founding narrative.[2] It's the first time this narrative has been brought into the public domain. We conducted high-end imaging of a particular rock art cave site, using lots of photogrammetry, time-lapse photography and gigapixel imaging, to enable those communities to tell this story.

It's been extremely well received and heralded as the 21st-century museology (Daley 2017). The exhibition will tour the world, which allows us to build a new dome. The overall exhibition at the *NMA* won the best exhibition in Australia for 2018 [at the Australian Museums & Galleries Association Awards].

KEIR: It seems to me that you're interested in exploring how these collaborations and technologies can create new knowledge, rather than seeing digital as a way of rethinking how old institutions run more efficiently? Is that correct?

SARAH: It's partly that. When I was at *Museum Victoria*, for instance, we talked of creating what I'm building now, but inside the museum so that there would be a more symbiotic relationship with the current processes for how they conceive design, think about, experiment, and implement these tools and technologies in association with their curatorial vision.

The biggest problem in museums, and it is beginning to change slightly, is the ghettoisation of digital into IT processes. That has limited the ability to think in a sophisticated way about how these technologies get implemented. Only a few museums have digital strategy and experience makers and enablers who are dealing with institutional-level frameworks. They're vital.

---

1 A fulldome system is an immersive, dome-based video projection environment.
2 *Songlines: Tracking the Seven Sisters* attempted to tell "an Indigenous founding narrative by using Indigenous ways of passing on knowledge. The project was inspired by an investigative collaboration between senior custodians of Martu country and Anangu Pitjantjajara Yankunytjatjara (APY) and Ngaanyatjarra lands of Australia's Central and Western deserts, along with the *National Museum of Australia*, the Australian National University and other partners." From the *National Museum of Australia*.

KEIR: I might be the choir on that.

MERETE: The way we work with digital at *SMK* is completely different. It's not about creating spectacular immersive experiences in a mixed physical/virtual atmosphere. I work mainly with leveraging the potential of spreading digitised content as widely and openly across the internet as possible so that we can reach out and become relevant for people who might never visit us in the physical space of the museum.

This is a completely different aspect of what digitisation and digital technology can do in a museum environment.

SARAH: Absolutely. At *EPFL,* we talk about "computational thinking" as a substrate for all pedagogy activities and research.

MERETE: For five years, I've been working with Wikimedia in Denmark and with colleague institutions to open up our digitised collections and collaborate with people outside our walls to enrich Wikipedia. This means sharing the knowledge and the digitised collections that we have and, in a dialogue-based movement, learning from the surrounding world about our collections – about what they can mean, the narratives they can foster and what they can be used for when we give up the control of how they used.

We believe that by opening up collections, and putting what is in the public domain in analogue form into the public domain in digital form, that we have a chance of making these assets relevant to new kinds of people and for new types of usage. Apart from it being an educational resource, it can also be a creative resource. We try to add a new level to our museum practice as the digitised collections become a toolbox for new creativity. We encourage and foster creative reuse.

Creative in this sense can both be the creativity that Wikimedians demonstrate when they use images from our collections to illustrate articles about topics where we would never had imagined those pictures would be used, and also hands-on creativity, such as clothes design, jewelry design, app development. Anything you can imagine.

KEIR: I think there's an argument that all museums, especially museums in the Western world, are challenged by the growth in the use of technology, both by the audiences and by their institutions. That demands huge resources. Sarah, you talked about a grant of $600,000 dollars, which is the operating expenses for a whole department, or possibly a whole museum.

SARAH: Exactly.

KEIR: With an initiative like *SMK Open,* or WikiLabs, you need the resources to digitise some or all of the collection, and at high-resolution to help compelling stories come to life or to make these resources available for creative reuse.

That is different to making sure the toilets are cleaned. That is different to making sure that the frontline staff are trained and ready to receive visitors. How do you generate the energy and the resources to make these kinds of experiences possible?

MERETE: That's an excellent question. These developments happen not because the resources and the strategy are in place from the outset, but because people within the museum respond to a development taking place around us – not only as museum professionals, but also citizens of the world – and feel an urge to respond. Then we start from the bottom up. The resource we use is our time and the energy that we can get away with dedicating to new things.

This way of working is exciting. It's also tough, and at a certain tipping point you need to turn this iterative bottom up, "start small, move fast" idea into something that becomes overall strategy of the museum. This is what we have done. You said, Keir, that in order to work with a DomeLab or with WikiLabs, you need a fully digitised collection with high-resolution material, but you *can* start with what you have. Even if you haven't digitised your entire collection, or if not all of it is in fantastic quality, you can start with the stuff that will work, and make the most of that.

When we started opening up our collection, it was a renegade project. We got permission to publish 160 high-resolution images in 2012. This was tiny. We have 260,000 art works in our collection, but 160 was all we needed to make the case. We did all kinds of crazy remix projects and started the WikiLabs and took part in *Gif It Up*[3] and everything we could imagine. A few years later, the base of evidence from that allowed us to write an open access policy for everything that's public domain in our collection.

KEIR: Was that intentional to do something that was iterative, building on the success, or was it more like a pilot to see how it went? Did you have a five- or 10-year plan in mind when you were working with the first 160 images, or was it "let's try it and see what happens"?

MERETE: It was in-between the two. We had a big vision, but we didn't have a carefully written five-year plan to roll it out. We worked at the challenge and tried to see how far we could get. What has been important is the huge generosity and knowledge network of practitioners in museums around the world, who we've been able to draw on. We're not alone in having this vision about openness and the potential of being relevant to more people.

KEIR: Was the 2009 *SMK* digital strategy, "We want to be a catalyst for user's creativity" the big vision? Did you publish the vision for the museum and your peers to see, or did you keep the vision as something that you were working towards through a set of experiences?

MERETE: We were very open towards trying out new things, but for a long time we didn't have much to show for it. We were inspired by the openGLAM movement.[4] We had the ambition to be a catalyst for users' creativity, so we

---

3 *Gif It Up* is a project run on and by Europeana that encourages creative reuse and remixing of cultural material.
4 OpenGLAM is a European Commission initiative, operated by the non-profit Open Knowledge Foundation, that promotes free and open access to digital cultural heritage held by Galleries, Libraries, Archives and Museums (GLAM) in Europe and the rest of the world.

put that in writing, and we had the support of a foundation to explore the potentials of digital technologies and user involvement at *SMK* from 2008 to 2012. The digital strategy you refer to was the formalised vision of that very iterative work that we did in those years.

KEIR: That's something that I've learned in my previous roles. If you can make clear a narrative with a destination, even if it's an exploratory destination, then doing pilots, doing experiments and iterating them within the framework of that larger narrative is productive in a bigger way than simply just doing the work.

SARAH: I was President of the Australasian Association for Digital Humanities [aaDH]. One of its many roles is to lobby the government to get resources directed to institutions and humanities researchers, including the GLAM sector, so that governments recognise that some of the resourcing issues that we're talking about are fundamental. I'm sure Europeana operates in a similar way as an advocacy platform for museums.

MERETE: Yes. There's a lot of duplicate effort happening where everybody is trying to figure out the same things in parallel. One of the important things that Europeana does is create shared common standards for the cultural heritage sector in terms of licensing, publishing frameworks, data models and engagement with diverse publics.

KEIR: Merete, do you consider one of your roles in this community as providing a tool chest for European organisations who are looking to do things more efficiently or successfully?

MERETE: Absolutely. We all face the same kinds of questions. We all follow the news and the development and it's baffling in so many ways. It's so rapid. It's difficult to keep pace, because the technological development doesn't take away the obligations we already have as a sector of collecting, cataloguing, preserving, et cetera. These things were difficult already, and now there's this new dimension, which holds the promise of making other parts of museum practice smoother, more flexible in the long-term. But wrapping our heads around how to use these technologies in a wise way is a big task for our profession. The kinds of expertise we use to hire are still important, but we also need new types of skills and expertise. And we need to add new skills to our portfolio in order to serve our audiences in a contemporary way.

KEIR: Sarah, you've done a lot of work in India and China. Could you explain who you're aiming to serve with these projects, and what learning and experiential outcomes you're hoping to achieve?

SARAH: Okay. I'll use Dunhuang caves project as an example (Kenderdine 2013). The Dunhuang world heritage site has 492 caves carved into an escarpment, with 45,000 square meters of mural paintings. It is the pinnacle of Chinese Buddhist art, but it is under significant pressure from visitors and climate change. Of the 492 caves, only 10 are open to the public at any one time. The Dunhuang Research Academy, who are the custodians, have 650 people working on this site, digitising their caves. That's a phenomenal number. At Pompeii, there are 10. We worked with the Academy to realise one of these caves digitally at one-to-one scale in a 360-degree system.

The original intention was to bring that cave to the Hong Kong public. For me, it had another context, which was the deployment of ultra-high-resolution data and 3D models inside completely immersive virtual environments, followed by a range of visualisation strategies for what you might do if you had 492 digital caves. It was a prototype for that.

KEIR: Wow.

SARAH: It cost $USD50,000 dollars. It was shown in China, in the *Smithsonian Institute*, all over the place. We then spun out five different versions. One for tracked head-mounted display where you could walk around inside the cave at one-to-one scale. One as AR application using 24 infrared cameras and one in a fully interactive fulldome for the World Economic Forum. These are all playing on the same idea in different modes. That was interesting from a data malleability point of view. The way you can transform data for many different devices and scale it across different machines and interfaces.

That became really interesting when juxtaposed with real objects. The philosopher Bruno Latour talks about a migration of aura from real objects to digital objects. That's also an interesting distinction to start to make between what is information and its delivery through digital tools, and what is an auratic artistic experience using those digital tools. Digital is often conceived as that augmentation layer with the info byte, as opposed to a fundamental experience, which has equal validity to the real object.

MERETE: A real breakthrough for our iterative work towards an open access policy here at *SMK* was when our young people's art labs collaborated with citizens living in Copenhagen next to metro construction sites. They did a huge remix of artworks from the *SMK* collection and mounted them on a 70-meter-long fence around the metro construction site. The people living there engaged in the remixing and felt empowered to turn this miserable construction process into a creative process and something they had some influence on aesthetically.

The breakthrough was when an esteemed art critic gave the project full coverage in the national newspaper and reviewed it as an artwork in its own right.

SARAH: That's a very important point. When the media endorses what are considered by many museums to be risky, unusual interventions, this has a very good flow on effect for future work.

MERETE: Yeah.

KEIR: What are the measures of success that you think of for these new models of working and these new sorts of experiential outcomes? Is it media? Is it foot traffic? Is it education and visual literacy? Is it the canon? Is it you learning how to do better next time around?

MERETE: For us, it's the social impact of being able to freely reuse digital cultural heritage. How does that empower people? How does it enable them to do new things? We are using Europeana's Impact Framework to capture the effects of *SMK Open*. The framework offers tailor-made methods for the cultural heritage sector to assess your impact. You can look back and see, did we have impact? Or you can plan for impact in specific areas. The methodology lets you look at cultural, social and economic innovation as fields where your work could have an impact, and also looks through different lenses such as learning, utility or community building. Europeana is proactively developing common standards for the sector because it's one of the places where we really struggle.

There's a deep internal conviction within the museum sector that our work matters to society, to economy, to people's lives. But how good are we at actually demonstrating that?

KEIR: That's a question for me, too.

SARAH: It's a huge question.

KEIR: As you answer, Sarah, could you think about how digital has transformed the world, and not just cultural experiences and cultural destinations? Technology has altered how humans experience the everyday and we can't be excepted from that. We can't be separated from . . .

SARAH: The post-digital.

KEIR: Exactly.

SARAH: The technology for me is completely irrelevant. It's about paradigms of experience and when we come to install these things in different communities, different communities are taking away different aspects of it, or contributing to different aspects of it.

KEIR: I believe documenting cultural practice or cultural heritage and making that data open is not a neutral act and may not always be a good thing. Releasing or presenting that material without context can be misleading or even harmful. Given you're depicting both tangible and intangible cultural heritage in these projects, how do you factor that into your decision making?

SARAH: It's certainly a big responsibility. We're doing a project on Hong Kong Kung Fu at the moment. We're bringing Kung Fu masters into the motion-capture studio and doing a lot of high-end documentation.[5] These are all living people. We had an instance where a contractor released some motion capture procedural modelling data on to the internet with no context, no reference to the masters themselves. That is very problematic. It needs to be guarded against.

I'm dealing in many cultures where they're giving me access to make high-resolution digital facsimiles of objects – whether they're 3D models or with images of highly prized material that will never leave the country – and I have to take care of that repository in securely and for the long-term. I'm making deals with universities for perpetual storage of this cultural material.

---

5 *Kung Fu Motion* was exhibited at ArtLab, *Ecole Polytechnique Fédérale de Lausanne* (EPFL), Lausanne, Switzerland, 2018/04/27–08/12. Collaborators: Jeffrey Shaw and Hing Chao.

In this era of heritage at risk, anyone can make a digital 3D models of polluted and destroyed sites. It's very common. There was one example in the U.K. where they crowdsourced the Temple of Baal images from Palmyra and created a digital model, which they then carved in Carrara marble at one-to-one scale and displayed it in Trafalgar Square.[6]

Critics look at this as a kind of massive cultural appropriation or neo-colonialism that the digital has unleashed in the documentation of other cultures, under a banner of heritage at risk. It's an extraordinary moment. It's not like iconoclasm is new, but now it's in the public domain as media spectacle. We are able to create phenomenal amounts of digital data, but it does come with this overtone of neo-colonialism and appropriation. Also, data hoarding. "Let's put it all away in a big storage facility somewhere and save it for the future." Or "mining patrimonial capital" as a business model. For many of these big data-gathering groups, that turns into a commercial outcome. It is important in this landscape to take care of the cultural sensitivities and proclivities when you're in the field.

MERETE: It's interesting how the power of the internet, of people being able to have a dialogue and work together and share information and knowledge and experiences with each other, is also raising these questions of loss of control. Are these good things used for bad purposes?

Any new technology is just a mirror of humankind. It's not especially new or surprising that these things coexist, but it's interesting how museums as places for learning and public engagement are caught between the positive and the negative forces at play here.

KEIR: I'm reminded of the Rijksmuseum's "adjustment of colonial terminology" project where they're replacing words in artwork titles and metadata to remove language that's offensive or derogatory or inappropriate.[7] Do projects like that have a ripple effect in organisations like *SMK*?

MERETE: We have the same process, and not just because of the Rijksmuseum. This is a transformation taking place all over the sector because post-colonial theory is becoming grounded in our practice. At *SMK*, we went through the curatorial titles given to artworks in our collection. Not original titles given by artists, but what curators historically had dubbed artworks that didn't have a specific title from the artist. We went through those with critical eyes and started updating exhibition labels so that we would no longer use the Danish equivalent of negro, but instead say, for instance, portrait of a man. This raised a huge debate in Danish media where right-wing politicians accused us of historical revisionism.

---

6 The replica of the Arch of Triumph was installed in Trafalgar Square from April 19–21, 2016.

7 Rijksmuseum's "adjustment of colonial terminology" project was introduced in 2015, removing racially charged and offensive language from digitised titles and descriptions in the museum's online collection. For instance, per Carey Dunne (Dunne 2016), the artwork now known as *Girl holding a fan* was previously described as *Girl from the Indies*, then *Eastern Type* and *Young Negro Girl*.

KEIR: Interesting.

MERETE: What we were really doing was just changing this information in the public galleries, but not in our database. All the original, historical data is kept, but we change how we name objects and interpret them in contemporary language. This is, to me, a completely obvious thing to do and something we do in aspects of museum practice all the time. We work with our tone of voice, we work with our communication tools and so on. This was just one aspect of that. It's politically loaded, but we were happy to have that dialogue with the public.

KEIR: In the article you published a little while ago, Merete, "Open Access Can Never Be Bad News" (Sanderhoff 2017), you make the point that by making your collection, your data and images open, you move them beyond the internal and privileged staff and scholarly use, out to where users are. That changes who gets to engage with cultural heritages, who gets to create and who has the agency to create.

I'm interested how that contrasts with the sort of contextualising that institutions do to make objects meaningful. Is it better to have more users, or fewer users who can really have a transformative experience with that work? How do we balance our educational goals with this open imperative?

MERETE: Let's see if I can balance answering all those really dense and important questions.

First of all, we need to realise that it's out of our control. As much as we, in our professional hearts, would love to control how people engage with our information, the content that we take care of and steward on behalf of the public, it's not possible. We can do everything in our power, but often we don't have the resources to make this perfect.

What we should consider is whether it's up to us to really make that value judgment. What is better? Is it the more deep, informed interaction, or is it the broader less controlled, diffused, out-of-our-hands interaction, which takes place when we let go of control like the Rijksmuseum has done?

I don't have a final answer for that. I know my own professional opinion, but I can't look into the future and see whether I was right. I just see that it's a losing battle if we want to control this development. We can try to work with it and influence it as much as possible in terms of investing in great data, in great ways of disseminating this content with the data intact, with clear frameworks for how the public can use and reuse these resources. Clear calls to action and ideas for meaningful ways of interacting with this. Yet even museums that really license and control their collections are not very successful.

SARAH: It was proved early on that nobody can really make money out of digital image licensing in a museum context.

MERETE: Exactly.

KEIR: It's been interesting to watch the *Metropolitan Museum of Art*'s justification for their Met Open Access project, which was both a gift to the community and a financial choice. I believe they were spending more on salaries to license images than they were making in licensing, so they transitioned to giving it away. For an institution of that size to be able to make a dual-pronged argument to the board and to the leadership – that this will save us money and give millions of people access to material that's hidden from them – is one of the keys for me.

MERETE: *The Met*'s collection is a brilliant example of material that has been collected from peoples all over the world and is now sitting there in New York City. Only the people who are privileged to go through the door have access. Even those people only have access to the fraction that's on display. With the open access program, this is, in practice, a way of handing back the ownership of this content to the peoples and cultures who first created the objects.

A lot of this material is in the public domain. It belongs to humankind. We need to keep in mind that we are trained to take care of these collections to the highest level we can, but we were not put on Earth to make a value judgment about what is a good way to use it.

KEIR: I've argued that putting material out on the internet without a proper description, without usable contextual data, is the digital equivalent of dumping paper files in the street with Polaroids poured on top (Winesmith 2017). Thinking of the incredible scale of the imagery that you're creating, Sarah, or the breadth of the stuff that you're doing, Merete, how do you make the content usable?

SARAH: For most of the types of material I'm working with, there are no standards. They haven't been developed. 3D intangible heritage data has no standard. Its interoperability, its usability, is still quite constrained. There are movements to expand the International Image Interoperability Framework (IIIF) image standards[8] into 3D formats. That means that this material can be released at high fidelity. I'm interested in high fidelity, not the "yellow milkmaid" image problem where there are thousands of versions of Johannes Vermeer's *The Milkmaid* on the internet and they all are completely different, and everybody has a different concept of what yellow is.

The integrity of data is really important to me. We need to be mindful of transitioning this material through time as much as we do our collections.

MERETE: How do we make it usable? The higher the quality, both of the imagery and the data, the more useful it will be to more diverse kinds of users and usages. I challenge that usage is always dependent on the context of the object. I can imagine someone out there running into a great image that inspires them to do something that wouldn't have happened otherwise without knowing the specific art historical context of the image.

We should be aware of and discuss the professional value judgements we were trained to care about. As an art historian, when I see a painting from the 19th century, all kinds of narratives are triggered. This is both a blessing and a curse. Other people without my training will see something different.

---

8 IIIF (International Image Interoperability Framework) a method for standardised delivery and description of digital images over the internet.

These artworks were not created for my profession. They were created for the diverse chaos of humanity. Having said that, I still believe that we should do everything in our power to ensure that the context is there if people want it.

[Due to local circumstances, Merete Sanderhoff had to leave the discussion at this moment. Sarah and Keir continue below.]

SARAH: Because I mainly do large-scale installation work, I'm often asked, "But why aren't the panoramas labeled? Why haven't you told us what site that is?" This is an important aspect of the way in which meaning in immersive experiences developed is emergent and interactive, and narratives are emergent because it's not predefined. It's up to the visitor to experience that place.

Context can be provided in such exhibitions, but there's no didactic messaging. Providing that juxtaposition between curatorial and these experiential modalities of digesting the world is important.

KEIR: Thinking about your projects and the way that they create a genuine sense of presence, different in many ways to the objects that surround them. Not everyone can build those. Not everyone has the skills or access to resources. What can you generalise out of that practice?

SARAH: It is about collaboration, of course. There's a quote from a curator at the *Smithsonian* from the 1890s, who said, "The future of the museum is where the library and the museum become a laboratory." Conceiving of one's space or position as an experimental laboratory, and finding those people that can help you build that is . . . These works are not born independently. You have to create visions for what you are doing. Most creative directors will endorse what you do if you can bring vision to the table. Find those avenues to bring to the table and find your best allies and collaborators either intra-institutionally or from outside of GLAMs.

Often this process for museums is cut short because of the tendering process and the over-reliance on the creative industries to provide technical solutions. I think this has been a travesty for museums. The project is delivered. The technology disappears a couple of years later because that was its intended life. There's not learning in the museum side. Because museum workers are no longer involved in the creation process, they don't gain any skills.

KEIR: One of the things that I would consider a great privilege of working in *SFMOMA* was that it was up to me and my colleagues to choose whether or not to go through a tender process or build things ourselves for many of our digital projects. When we used vendors, we created contracts where the IP and code was handed over and we ran the projects out into the future.

SARAH: These are important issues. The ability to design and create and curate at the same time is hugely beneficial and grows the internal robustness. There are economic arguments to be made, too. The re-centralisation of those creative processes is fundamental to the health of an organisation.

Having given up my museum job after my entire career in museums, I find myself in a very different context. Sometimes it feels pretty wide open. I can go anywhere and do anything with lots of powerful resources. Bringing this all back together in a new context allows me to rethink what I'm doing. Mainly this is around moving image archives because they're often the archives that you can't distribute on the internet or easily due to copyright restrictions. Working on situated engagement using high-end computer vision and machine-learning tools is where I'll be working with some of the major archives of the world to solve that problem for museums. These are collaborative projects. The hybridity of these roles enables me to take up these opportunities and to shift my work in ways that I want it to go. Certainly, the ability to traverse many organisations at once is kind of vital to the multidimensional facet of any project. Most curators do this anyway. They're working at multiple levels with multiple communities and stakeholders and collaborators.

KEIR: Could you speculate on how visitors and users – how museums – will change how they present digital heritage, or even heritage at all, in the next 20 years?

SARAH: The term user has been used a lot in media arts since the 70s. Another word for it, which I prefer, is participant. A participant in an emergent experience. One can intuit from a big interactive digital installation, with multiple types of data that you make emergent narrative in, the way in which the museum may be interacted with. There are no set curatorial narratives. Everybody comes out with a unique experience.

It's not that this is a new concept, but the ability to make narratives across museum collections from one museum to another while you happen to be visiting one museum is to make the object much richer than it is in its current context. Many people talk about museums' objects. By the time they enter the museum, they are denuded of resonance and how you re-create that is an interesting proposition because it relates to an object as heavily loaded with its whole history of its creation. Every object is an assemblage of many narratives and how you can use these to connect between different organisations.

It's the connecting of silos, and that emergent narrative that might come out of that, which is what I'm interested in. It's what historians do. They look at specific archives and they piece them together to create these narratives, but we can do it at a very rich experiential virtual digital level potentially. That is one thing I think about quite a lot.

KEIR: Thank you, Sarah.

This conversation reveals a few tensions related to digital cultural collections, colonialism and open access worth unpacking further. Over the last 15 years, museums have increasingly made parts of their collections available online for access and, in some cases, reuse by the public. As open access to collections has become more commonplace, its nuances have become more visible. For instance, there is a tension between the desire for openness, which seeks to make digitised culture freely available for reuse by anyone, for any purpose, and the need to restrict access to certain kinds of material, such as the motion captures of the

Kung Fu masters, or to give more nuanced access to Indigenous collections. As Keir notes, "documenting cultural practice or cultural heritage and making that data open is not a neutral act and may not always be a good thing."

At the heart of this tension is a question about who has the authority to decide whether digital information about museum collections should be open, to whom and under what circumstances. As Mathilde Pavis and Andrea Wallace write in their "Response to the 2018 Sarr-Savoy Report," "The management of intellectual property is a cultural and curatorial prerogative, as is the initial decision about whether and what materials to digitize. These prerogatives should belong to the communities of origin" (Pavis and Wallace 2019). Working responsibly with communities in the digital age means enabling individuals and communities to have agency in how their stories and histories are presented, represented, and used by museums and their publics alike.

For more on open access and the free use of cultural resources, we suggest you read:

Conversation 6: Kati Price + Loic Tallon
Conversation 7: Shelley Bernstein + Seb Chan

For more on digital colonialism and heritage at risk, we suggest you read:

Conversation 2: LaToya Devezin + Barbara Makuati-Afitu
Conversation 3: Lara Day + David Smith
Conversation 4: Sarah Brin + Adriel Luis

References

Daley, Paul. 2017. "Songlines at the NMA: A Breathtaking Triumph of 21st Century Museology." The Guardian, September 15, 2017.

Dunne, Carey. 2016. "Danish National Gallery Removes the Word 'Negro' from 13 Artworks' Titles." Hyperallergic. 2016. https://hyperallergic.com/304385/danish-national-gallery-removes-the-word-negro-from-13-artworks-titles/.

Kenderdine, Sarah. 2013. "'Pure Land': Inhabiting the Mogao Caves at Dunhuang." *Curator: The Museum Journal* 56 (2): 199–218. https://onlinelibrary.wiley.com/doi/abs/10.1111/cura.12020.

Pavis, Mathilde, and Andrea Wallace. 2019. "Response to the 2018 Sarr-Savoy Report: Statement on Intellectual Property Rights and Open Access Relevant to the Digitization and Restitution of African Cultural Heritage and Associated Materials," March. http://web.archive.org/web/20190610002217/**https://zenodo.org/record/2620597**

Sanderhoff, Merete. 2017. "Open Access Can Never Be Bad News." SMK Open. 2017. https://web.archive.org/web/20190712193610/**https://medium.com/smk-open/open-access-can-never-be-bad-news-d33336aad382**.

Winesmith, Keir. 2017. "Against Linked Open Data." Medium.com. 2017. http://web.archive.org/web/20170630234622/**https://medium.com/@drkeir/against-linked-open-data-502a53b62fb**

CONVERSATION 6

Kati Price + Loic Tallon

". . . the two biggest barriers we face working in digital is that we're dealing with ambiguity and we're dealing with change. And those are two things that humans tend to hate."

– *Kati Price*

FIGURE 6.1

Loic Tallon (U.K./U.S.) has been a central figure in the museum technology space since he began organising an online conference series on mobile technology in museums in 2009. He later joined *The Metropolitan Museum of Art* as a mobile manager and rose through the ranks to become their Chief Digital Officer. While at *The Met*, Loic published a series of articles on their digital structure and future.[1]

Within a few months of Loic's article, Kati Price (U.K.) published and presented a comprehensive survey of how museums resource, fund, and structure their digital work and their digital departments, co-developed with Dafydd James of *Amgueddfa Cymru, the National Museum Wales* (Price and James 2018). Producing and sharing this sector analysis and insight built on Kati's long track record of publishing the ideas and practices behind the V&A's own digital output.[2]

Keir contributed details of *SFMOMA*'s digital approach, responsibilities, priorities and team structure to Kati and Dafydd for their research, and to Loic for his, and was confident that bringing Kati and Loic together to explore digital practice and team structure at scale would lead to a rich and informative discussion.

Alongside millions of locals and tourists, we try to visit the *V&A* and *The Met* whenever in London or New York. The scale, richness and variety of their respective collections are phenomenal, providing incredible material for their digital teams to respond to.

This conversation was recorded on January 9, 2018. At the time, Kati Price was Head of Digital Media & Publishing at the *Victoria & Albert Museum,* London, U.K. and Loic Tallon was Chief Digital Officer at *The Metropolitan Museum of Art,* New York, U.S.

KEIR: Today we'll be discussing digital practice at two of the largest museums in the world. I'd like to start today by exploring the strategies you apply to manage the complexity of working in a multi-faceted, multi-layered institution.

How do you structure your teams?

LOIC: At the highest level, our organisational structure follows our business goals and workflows. As these goals and workflows evolve, we try to responsively tweak and adapt our organisational structure so that we are working efficiently and with clearly defined decision-making responsibilities. In my time at *The Met*, we've probably had half a dozen such organisational

---

[1] The articles were published on *The Metropolitan Museum*'s website in their blog section. https://www.metmuseum.org/blogs/listing?&facetname=Loic+Tallon&facettype=author
[2] The articles were published on the *V&A*'s online blog. See https://www.vam.ac.uk/blog/author/kati-price

changes, the most impactful of which was in late 2016 when we introduced product management, UX design and agile into the museum's digital practices. This was a significant rethinking of how we were working. It importantly aligned our practices more closely with the digital sector as a whole.

What is quite unique at *The Met* is that the digital department is responsible for not only the public-facing digital presence of *The Met*, but also the management of the collections data: this data provides the platform for almost every one of the 31 million engagements on the website each year, and 100+ million interactions on third-party platforms. The business goals of the team responsible for managing the collections data – the collections information team – are different from those of our public-facing digital teams, but by having them sit within the same department, it enables us to ensure that the way we clean, manage, structure and license that data has an eye to how it will be used by our audiences and not just by our internal stakeholders.

Of the 65-staff in the department, the majority – around 50 – work on our public-facing digital activities. This covers both content production and product development, and their responsibilities are broadly split between two high-level goals: contributing to a world-class on-site visitor experience of the museum across our three NYC venues, and extending the reach of *The Met* to a global audience who will likely never visit the museum. There are separate KPIs for each goal, and so we've structured the teams into squads based on those goals. We call it Path 1 and Path 2.

Each team works slightly differently. Our product development team work 100% agile, scrum methodology.[3] They're incredibly disciplined about the agile process they follow to deliver value. Decisions are made on a bi-weekly basis about what to work on next to best meet our business goals.

You compare that to the collections information team, where most of their work is ongoing maintenance and management of the data. Changes to data fields, supporting the upload of a batch of images or cleaning up the data on older records.

Increasingly, we're directing our stakeholders to contact the team they want to talk to, rather than just the "digital department." The word "digital" is relatively meaningless. We want to help our stakeholders better understand exactly what they're asking for. Knowing which team to talk to is part of that. Is it a content question you have? Or is it a product question,

---

3 Agile and Scrum encapsulate a set of related practices for project management that encourage an iterative approach to product development. Although most commonly and successfully applied in the development of technology, these approaches are increasingly being effectively applied in other sectors. As with most project management methodologies, it is not suited to all circumstances nor consistently utilised when suitable.

for example a new feature or functionality? Is it for the on-site experience, or is this for global audiences?

What stitches this structure together is the idea that it is our responsibility, as the department, to drive the digitisation of *The Met*'s practices and programs in order to best fulfil the mission. I believe that's the larger purpose of the department.

KEIR: Kati, how does that balance against what you do at the *V&A*?

KATI: I have a much smaller team than Loic. We cover both digital media and publishing. We have a team of six making books and catalogues, and managing third-party relationships with publishers. On the digital side, there are 14 people, split across content, technology and design.

Technology and design work as a product team. Thinking about products is a shift away from a world where digital investments were seen as projects with a finite end. For example, with a new website, the launch is just the beginning of the product lifecycle. A product culture helps instill a sense of product ownership and requires a vision for those products based on user research, data and business goals. It leads to better understanding of the connections between the different products. Ultimately, having a product culture helps you plan and prioritise, and make visible the variety of work that a digital team undertakes.

We're always hiring for particular skills in the product and content teams. But alongside technical, design or editorial skills, I focus on attributes like resilience, curiosity and excellent communication skills. Resilience is necessary. Curiosity – which helps uncover new processes, technologies, trends and so on – is common in the types of people that are attracted to working in institutions like ours. People who are curious tend to make better work. And then, communication skills, without which we can't do our jobs. We need people in digital who are extremely articulate about what they do, why they do it and how they do it. It's important they don't use jargon.

We work in a very agile way. That's sometimes at odds with the pace and the processes within the museum, but it's increasingly understood and accepted within the organisation – even informing the way we do non-digital work. Within the digital team, we finish each sprint with a showcase, so that anyone from the organisation can see what content we've produced that fortnight, what books we've published, what digital development has taken place and what design work we've delivered. The product culture enables greater transparency and visibility of what we do, and a chance for the organisation to understand and steer the activity that we oversee.

KEIR: Do people come to your demos?

KATI: Most people want to be involved with the digital agenda but can't always make time to do so. Having a dedicated slot for showcases [means] they can check in if they have time and inclination.

It's interesting to see how the make-up of the audience changes. At the moment, we're working on site search, because we launched our website without search functionality. We've now got key stakeholders coming to our showcases that haven't previously. And, if we're working on retail product, we might have lots of the retail team join.

They [the showcases] are very informal. We had internal team showcases before, but when we were doing our major overhaul of the website, we opened them up, and ended up holding them in the *V&A* lecture theatre. It is a deeply intimidating space and quite the opposite of the sort of informal place you might ordinarily have a demo or a showcase. But it lent importance and weight to the project, and gave it a lot of visibility. We'd have a packed-out theatre, with lots and lots of questions from the audience. It was a great way to get feedback, and let people feel like they were feeding into the project. Most importantly, it meant there were no surprises for our stakeholders later down the line. Working within a product culture like that, you don't have to have committees signing off on things, because people have been part of a process all the way along. They know what they're going to get.

People across the museum are now using showcases or "spotlights" to provide a window into what they do. Teams that don't necessarily use agile methodologies are using some of the agile rituals to help bring that transparency and consequent staff engagement into their work. That's been an interesting side effect that we hadn't anticipated.

KEIR: Kati, it seems that digital departments are often situated due to personalities, happenstance, or accidents of history, rather than strong intent to say, "We're going to have digital and we're going to place it here." I know you and Dafydd James have done some research into this (Price and James 2018).[4]

Thinking about your comment about taking this agile showcase methodology and syndicating it, like seeds flowering throughout the organisation, I don't often see that openness to digital practice permeating in other institutions. What do you see are the emergent practices about how to effectively situate digital-informed practices and thinking, and the digital department itself, in an art or design museum?

KATI: I think they're two very different things – situating the team and situating the thinking. The most important thing is that your digital team's work is aligned with the organisational mission. To a large extent, it depends on what your organisation's aiming to achieve more broadly, and how digital might support that – that gives clues as to where it's best situated.

For lots of smaller organisations, engagement and access are seen as key drivers and, therefore, digital is often seen as a function of the marketing team. For others thinking about digital in the context of business

---

4 Their research chronicles digital team structure, staff competency, internal remit, staff renumeration, team size, departmental expenditure and other important factors across 56 institutions from the U.K., U.S., Canada, Europe, Australia and Brazil. We encourage you to digest and then contrast their findings with the ideas covered in this chapter.

transformation, you wouldn't situate a digital team within a marketing team, because you're not going to achieve broader-scale impact.

And in the case of business or digital transformation, where the organisation has a goal of disrupting how the organisation operates, you are necessarily going to have digital reporting to the leadership level of the organisation.

I don't think there are right or wrong approaches, although some people Dafydd James and I surveyed reacted quite strongly against the notion of digital reporting to marketing. In the U.K., a couple of the national museums have marketing reporting into digital, and I think that's a really interesting move. It's not to undermine the value and impact digital (or any other) marketing has, but it's just one area that digital technologies and activity can support.

KEIR: I'm excited to see for the first time a few people leading organisations, mostly in regional or smaller institutions, who have come through the digital track.

Loic, my read of your recent writing is that you believe that a centralised digital department adds value and is important, but that the skills and practices need to be distributed and deputised (Tallon 2017). Is that right?

LOIC: I don't believe in centralising all digital skills, and that does call into question the role of a digital department. I want to move the discussion to where it does not matter if people report to a digital department or not. That's not actually the question. Is there a better way of doing what you're doing using digital? And if there is, we'll help you do it.

Whether it is a curatorial piece around a collection, whether it's about how we make revenue, whether it's about how we capture emails, how we build community or how we define what membership means. All of those questions, I think, are ones where the advent of digital technology has had a significant impact in answering and improving.

One commonality most digital departments have is product. And I think that when many people use the word digital, they actually mean product. There's no digital department I'm aware that doesn't have product.

KATI: Before doing our survey, I would have agreed. We surveyed quite broad types of institutions. Of the 56 we surveyed, half were museums. The rest were libraries, archives and galleries, and some were really small, with under five people in the digital team or managing digital activity. To some of those organisations, talking about product and product culture would still be unfamiliar territory. If you're talking about product, it's likely you have a number of set skills in-house, namely technologists, people building software; designers, who will work across UX, interaction design and visual design; and product management. Lots of organisations are so small that they don't have that, and they're never going to have that.

Where there is commonality is around content. We found that if you stripped out everything else, content was the one thing that everybody did. The focus on product is probably much more relevant for larger institutions like us, where we've got lots of in-house expertise. That isn't so relevant at the smaller institutions.

LOIC: That's interesting. For the smaller institutions that are outsourcing the websites, they're still effectively doing product management in-house. The person managing the website is basically a product manager, whether they're aware of it or not.

KATI: I agree with you. Outsourcing a website rebuild project doesn't mean they're not managing product. But manifestations of a product culture, things like visible backlogs and prioritised user stories, might not reside internally in the institution. They might be held externally by an agency, for example.

Although there's never going to be the perfect digital maturity model, I think we are aiming collectively towards a place where digital becomes distributed across the organisation. However, I think in large organisations like ours, you will always have a kernel, a core of product teams, so there will never be a truly distributed model.

KEIR: Can I argue against that as a nirvana? One of the things that I was first attracted to about museums is that a museum is a bastion of specificity. There's someone at *SFMOMA* whose job is to conserve objects made of one set of materials. They are the best steel person I've ever met. There's no distribution of that understanding of collecting and maintaining an artwork for hundreds of years. How do you keep a Richard Serra sculpture the same timbre of burnt orange for 200 years?

The places that I see that are most progressive are the ones that have taken digital product making and digital practice and syndicated that around the museum –whether that's how design requests are lodged or using an iterative showcase model. I'm curious about the non-digital arenas, and where you see the potency for those practices that you both champion across the organisation. Do you think there are transformational opportunities there?

LOIC: The transformation needs to be in the culture. The showcases Kati references are important examples of new behaviours that are being adopted by the institution. If you look at the kinds of behaviour and work culture the institution needs to adopt in order to succeed with digital, these are not behaviours that are native to the institution of the last 150 years.

To illustrate this, let's look at the process of designing an exhibition. Traditionally, an institution spends 100% of the exhibition budget prior to opening so that it is "perfect" on opening day, for the press preview. If our goal is to deliver the best exhibition for our visitors, we could arguably deliver better results spending 80% of the budget before opening, and then, for all successive weeks, ask visitors how they're enjoying their exhibition and get feedback, and update the exhibition once a week based on that feedback. The exhibition would be better on the day of closing than on the day of opening. I'm sure this type of approach would meet resistance internally, but that's basically what we're doing with our digital products . . . we want to launch the smallest thing possible to test the idea and start getting user feedback.

When I think of digital transformation, or business transformation through digital, that culture change component is really important. If I was to say the nirvana piece of a distributed model, it would be understanding why working in a particular way in today's business world is more beneficial than the way we worked previously.

KATI: Totally. Nirvana isn't having product culture; the nirvana is a distributed system.

It's interesting thinking about how some of the principles we apply in digital development are expanding into the broader museum and its practices. Take user-centred design processes, being utterly driven by the needs of the user, but being mindful of business goals at the same time.[5] Thinking of the physical building in user-experience terms is an interesting development. The fact that you see the rise of the Chief Experience Officer, rather than the Chief Digital Officer.

What's going to be really interesting is seeing where digital ends up, in terms of leadership roles.

KEIR: What do you do when senior staff or other departments push back against this sort of practice?

KATI: Transparency is key. If people can see what you're doing and understand why you're doing it, and the tools you're using to do it, it helps demystify this thing that is digital. Having an appetite for understanding how digital media and technologies can improve their working practice . . . we're up for conversations about that. But I don't think everybody across the organisation needs to be onboard with some of the more nuanced aspects of digital development. You know, we've got amazing subject experts here on the curatorial or conservational staff. They don't need to know the ins and outs of our job. Why would I expect them to?

If people are resistant, understanding their goals and demonstrating how you can help them achieve their goals is a great way of showing what can be done.

KEIR: Yeah. I guess what I'm interested in are strategies to break through executive or board inertia, or institutional fear of working in a different way.

Even though *SFMOMA* is in Silicon Valley, and the board understands what the digital department is trying to do, there are members of the executive who are uncomfortable with the idea that we don't know exactly where we're going to get when using an iterative, public, transparent design process. I've often bashed into roadblocks where the approach we're taking is

---

5 In this context, *user-centred design* referred to the contemporary product, interface or experience design practice that preferences user needs and expectations over business imperatives. This approach typically including testing with real world users – the outcomes of which lead to changes in the product or experience.

being questioned, let alone the expense and the audience, or a perceived drift from a very traditional mission.

LOIC: That speaks to what you said earlier around seeing colleagues more from the digital vertical ending up in leadership positions. Because the culture and behaviours these colleagues are bringing are to their new senior roles is enabling its adoption in new parts of the organisation. I think, over time, you'll see more people in those kinds of positions. If you were to look at a number of the emerging leaders inside museums, they've come up quite often in the digital space, and these ways of working are native to them. And they're delivering results.

To your question about how we bring people with us, number one is to focus on the mission. Focus on the mission relentlessly, and point out that what you're trying to do is fulfil the mission in the best possible way we are able to. I've always found that conversation to be fruitful.

I also love the idea of finding one of the longest-serving "institutionalised" colleagues in the museum who agrees with the way digital is working, and position them to advocate for digital on your behalf. Because then we can move past this ageist assumption that "young people" see the world differently from "older people." I occasionally hear that it's an age thing. It's not. It's just how people approach their work.

KATI: I think the two biggest barriers we face working in digital is that we're dealing with ambiguity and we're dealing with change. And those are two things that humans tend to hate. It's not to do with age, it's to do with being human.

Lots of people are averse to those two things. Where you find resistance, it's where people don't like the ambiguity of not knowing what you're going to deliver at the end of a project. It can be mistaken for you not knowing what you're doing, rather than accepting you're using an iterative, user-centred approach.

Change is going to be a key part of this. And lots of people are happy to get involved in change, but they [don't] like change being done *to* them. A lot of it is around how you manage their role within a broader piece of business transformation, making them feel part of it.

The way we developed our business case for major investment in our digital infrastructure was to show what could be done with a few small, tactical interventions. And say, "Look, here you go. We added this new feature for our newsletter sign-up. X amount more people signing up to the newsletter means Y amount more ticket sales or merchandise sales," et cetera. "Imagine scaling that up."

It's important to think of ways to demonstrate that this isn't just about making new things for the sake of it – that this is actually about driving our business goals. And showing in very small ways how they can be achieved by doing at a smaller scale some of the things you anticipate with broader, larger-scale investment. That's how we got people onboard, because we were able to show them what it might look like, and show them the kind of results they might expect.

LOIC: I think we've got to acknowledge that we're usually not the first person doing digital in our institution. And recognising that digital itself is only in its "teenage" years. We haven't matured as a sector yet to say "these are the clear directions" . . . there's no set of established best practices, per se. We need to recognise that this is different from most other departments in the institution, and will naturally create a sense of caution from leadership.

Most institutions have probably now spent a fair amount of money doing things over the last 10 to 15 years, and aspects of that investment are probably deemed to have been successful. Others not. If I was to look at the investment *The Met* has made in digital over the years, and how much of that still has value for us, I can understand why some executives will raise their eyebrows. I completely understand why certain executives, certain trustees, will say, "Hey, we're spending all this money, and we've been doing it for the last 10 years. What have we got to show for it?" Acknowledging that risk is really important. We need to be ready to answer those questions.

KATI: Thinking about the bids I've made for digital investment, you need to make a decision to take one of two approaches: Are you going to go down a reputational route – we have to do this otherwise it'll look bad – an ageing website, or antiquated UX? Or will you take a more commercial approach – for the digital investment to deliver ROI? If you're going to demonstrate ROI, you can start saying, "Look, we do this, and we'll generate X amount more memberships, more donations, more merchandise sales," et cetera. But that there are lots of areas of the museum that can't demonstrate ROI, and neither should they.

Digital isn't unique in that sense, because people are having squabbles about investments all over the place. We are competing in a high-pressure environment, where we don't have as much budget as we'd like. We all have massive ambitions that are bigger than our ability to serve on them, so we have to make those decisions on priority areas to invest in. And digital is up for discussion just as much as any other area of the business, including the building.

In making the case for digital investment, I find that it can be very useful to use analogies from the physical environment. At the *V&A*, we've got a courtyard in the middle, and we re-turf the grass every three months. And, at the moment, it looks shocking. It's been raining and kids have been trampling over it over school break. There will be internal meetings where they're debating over when we need to re-turf. And do we need to re-turf as often as we do? And could we be re-turfing in different ways? But that's part of the business of keeping the organisation up and running. We need to have a presentable and enjoyable environment. Digital is no different.

KEIR: A huge part of roles like ours is to constantly be optimistic, and to be good narrators – to be people who can tell the story of why the change is needed.

Because, in the end, it's those stories that are balanced against other stories, like frequency of turf replacement or conservation of material that may not go on view for 10 years. If you can't articulate the need, can't articulate the audience, you can't articulate the reputational impact, then you won't be at the table when it's time to order the museum's priorities.

To change gears a bit and talk about audience, *The Metropolitan Museum of Art* in New York and the *Victoria and Albert Museum* in London are two of the largest and most visited museums in the world, situated in two of the most vibrant and diverse cities in the Western world. When I think of them, I get a mental picture their impressive buildings, however both organisations are physically distributed and well as digital distributed. Kati, what's *V&A*'s role in London?

KATI: When you're saying the *V&A*, I guess you're talking about the building I'm in right now in South Kensington. But the *V&A*'s a global brand, and you can visit a *V&A* gallery now in Shenzhen, in China. You can visit the *Museum of Childhood* in East London. Later in 2018, you'll be able to go to the *V&A* in Dundee. After that, we'll be opening up two new sites in East London. So, it's an interesting question and one that I think the answer to which will shift over the next few months and years.

Wherever we are, we promote knowledge and research about design, and the designed world. We have this fantastic collection of objects, and each of them can tell us many different stories about our designed world and what matters to us, as individuals and in broader society today. I think that's what keeps people coming back and gives us a unique position to talk about what's happening in a global context.

So, we have things like Rapid Response Collecting, which is very much aimed at questioning the relevance of particular contemporary objects – the truths they reveal about how we live today. We're able to raise questions about economic, political and social change, globalisation, technology and the law. For example, the implications of the manufacturing process behind Katy Perry eyelashes – the fact that the women making these eyelashes are going blind by the age of 40. We start having conversations that are much bigger than the relationship of the *V&A* to its immediate geography.

KEIR: Did you acquire Edward Snowden's laptop for the rapid response collection?

KATI: Yeah. The hard drive.

We increasingly collect those objects that are digital in origin. We're starting to talk about objects that have digital DNA, and thinking about the systems and the infrastructure we have – and will need – to collect and conserve those objects. We added WeChat, China's largest social media platform, to our permanent collection (Cormier 2017). How do we manage, preserve and enable access to that digital object in years to come? We've got a research project looking at that, because the objects we're collecting now are more complex and challenging in many ways, because of their digital DNA, than the objects from our past.

LOIC: Those are awesome objects to collect. The *V&A* acquired the model, I think, of a 3D-printed gun, didn't you?

KATI: Yes. With some very deliberate revisions made to it, so that it can't be fired. The *Science Museum* over the road also collected that, but the story about the *V&A* collecting that object is a very different one to the *Science Museum*. The *V&A*'s take on an object is about humans and our relationship to our designed world, the decision we, as individuals make within it. It's a different story to the science side of things.

KEIR: Can I extrapolate that you would consider the online audience anyone interested in the designed world? Is it as broad as that? Or is it like anyone with an internet browser, when you consider your online audience?

LOIC: I definitely want to encourage institutions like *The Met* to think big. At *The Met*, we are privileged to have a collection of 1.5 million artworks, covering 5,000 years of world history. There are 3.9 billion internet-connected people in the world. Our aim is to reduce the distance between the person and the objects that inspire them. And whether that's a question of access, language, platforms, interests, whatever it might be, our aim is to reduce that distance. So, I think of our audience as every single internet-connected person in the world.

It's really important to make the distinction that I don't expect all those people to come to our website. Our website will play a core role – we have 31 million people coming to our website. But I recently wrote about our one-year anniversary of open access at the museum, and the impact of our relationship with Wikipedia. One of the examples I love is that we have Henry VIII's armour, which is viewed 180 times a month on our website. But the Wikipedia entry on Henry VIII is viewed 405,000 times a month, and *The Met*'s object is part of that article. And whilst the experience of the armour is very different within the context of a larger Wikipedia article, and as one of 22 images on that page, the access to the artwork in that context is really important. It's doing it in ways, in languages, in contexts which we cannot replicated internally.

So, I definitely think about audiences as anyone who would have an interest in the collection. And I believe everyone in the world would have interest in at least one object here.

KEIR: I was looking at some numbers and there's about twice as many references to *The Met*'s artwork on Pinterest as there is on *The Met* website. So, that's a commercial entity that has advertising, that may or may not disappear, that may or may not have the same goals and aspirations as *The Met* does.

Partnering with Wikipedia feels like a transparent good, but what about partnering with Google, Pinterest, Facebook or places where it is not actually obvious that these organisations have the public's good in mind? Is that a discussion internally? I struggle to find the right way to work with some of these corporate brands, because the things that they measure as ROI and what our institution does, the values are so different that I often can't find commonality.

LOIC: Yes, this is a big discussion internally. We believe there is significant value in collaborating with these types of partners, but we need to go into it with our eyes open, and with a firm awareness of our partners business model and goals. If we do that, I think we can find areas of mutual interest.

I don't think that was always our approach though. In the past, too often it has been a major technology player approaching a museum and saying, "Hey, we're thinking of doing this." And we say yes because, oh, it's Google coming to talk to us and this makes us sound hip and feel special. Whereas, I think we can switch the conversation around and say, "This is what we're trying to achieve right now. These are our business goals. How do they align with what you're trying to achieve? Is there alignment there? Is this worth working on it together?"

So, whilst I agree that the relationship with Wikipedia is a natural one, it's an obvious one, I genuinely believe we can build relationships with other commercial technology partners. I mean, Pinterest is the number-one referral of traffic after Google searches to the collection online. It's a huge draw. Why not work together to amplify that?

I think it would distance us from our users if we placed a restriction on our ability to partner with organisations based on them being commercial. We need to be looking where our users are. What platforms are out there? And where can we put our collection, the objects, in the places that are most relevant to them? In many cases, users have adopted commercial platforms to meet these needs.

KATI: I agree with that. I think one of the biggest challenges for us is data silos. Lots of the sites you've mentioned are data silos, which means they're essentially like little snapshots of what we knew about an object at a certain moment in time. As our knowledge changes and increases, and improves, that data will stay as it was at the time of that partnership being signed, or at the time somebody took something from one place and popped it on a Pinterest board. It won't be connected to our core systems.

How can we ensure that the data is not kept in silos, but instead is constantly updated? That's what's best for the institution as well as the user.

LOIC: I 100% agree that the data has to be kept up-to-date. When I think of the partners we're talking to, that's now a precondition. Just to be clear, Wikipedia don't do that yet, so what is presently live is a data dump that needs maintenance on an ongoing basis. But with Wikipedia, we are working with them to help drive the types of changes that would maintain the integrity of the data and have long-term value for us . . .

KATI: Value not just for them, but for us and for the user. Everybody in that context.

LOIC: . . . We're not going to play unless these following conditions are met.

The stuff with the Wikimedia Foundation that interests me most is actually around Wikidata, which I hope will become a framework we can all start building on. You think of some of the projects museums have tried to do about connecting datasets across institutions, so that the Pablo Picassos in

The Met are linked with the Pablo Picassos at *Tate*, and the Pablo Picassos at the *V&A*. On our scale, it's very tough to replicate that.

But on Wikipedia, there's a whole community of people who are doing this already. And I think Wikidata provides a potential framework to start integrating with that. I feel there's something worth exploring. It's an area we're really putting energy in and time into right now.

KEIR: It seems a Herculean feat of institutional bureaucratic jujitsu to get *The Met*'s open access policy to fruition, to get almost 400,000 images out into the world, to remove the copyright restrictions and all of that. I think it's incredible and lauded appropriately.

I've argued repeatedly that publishing data is not necessarily good, and "I can't wait to see what people do with it" is a terrible ROI. Whereas, it sounds to me like your study projected genuine ROI, about changing how the internet is consumed to make it more accurate, timely and with better authority. What does that mean for the institution? How does it learn? How do the staff practice this change by what open access creates?

LOIC: *The Met*'s mission is to collect, present, preserve and study works of art in order to connect people with knowledge, ideas and creativity. We're doing that on Wikipedia already, so we have mission fulfillment. We firmly believe that.

Controlling knowledge doesn't really exist. I mean, we'd lost control of the images before we went CC0. We're just acknowledging it. And saying, "okay, if we are losing control, let's enable people to use this content in a way that suits their needs, rather than hang on to this notion of control."

It's a culture change internally. We need to reach people where they are. So, how can we then leverage technologies that are out there to do that for us?

KEIR: Are there things we can do in the digital space that are not possible in any other context?

KATI: One project we're doing at the moment is Immersive Dickens, which takes one challenging object – an original manuscript by Charles Dickens – and creates an immersive experience around it. Whilst it will have a physical manifestation, it's drawing from user-centred methodologies and incorporating immersive sound technology (V&A Research 2018).

An MVP [minimum viable product] for an exhibition would be centred on just one object.[6] It's challenging to create an immersive experience around one object, but it poses exciting opportunities around how we might develop exhibitions and displays in the future. We're unpacking what we mean by immersive experience. We're looking at how to use

---

6 A minimum viable product (MVP) is a product with just enough features to function and be usable by a subset of the intended audience. Beginning an MVP allows its creators to learn from its use and improve as they go, rather than launching with a full-featured product that is expensive to produce and may not be what its intended market wants. This term was popularised by the software development community.

technology as a way to create that immersive experience, but by focussing more on our users – in this case teenagers studying Dickens – in the narrative development.

KEIR: Loic?

LOIC: We haven't got a project where we dedicate ourselves to one object and dive deep. I probably get too quickly concerned about questions of scalability.

On our website right now, if you click the highlights button, we're at 2,000 objects. I'm negotiating that down to 470 objects on which we have foreign language content. Those 470 objects are part of a guidebook that's been translated into 10 languages. And we're in the process of publishing that content online, so we've got to create a 10-language experience on the museum's highlights in a moment in time (the book was written eight years ago).

Interesting for us to compare, Kati, what you're doing, which in one sense sounded much more compelling and engaging. We're still in that space of creating rules that we can apply to large numbers of objects, where I can say our highlights have got X amount of content or stories created about them for X types of audience. To get certain rules like that, which I know I can scale with.

KEIR: A lot of what we talked about is bringing contemporary best practice from the field of product-making or the field of media into the museum context. This openness to experiment, to give things away, to be generous, to be going to where people are is emblematic of the last 10 years.

Flipping that over, what would a world look like in the future where museums led the way in how content was created, content was disseminated and stories were understood? What does that world look like? Is that even meaningful or helpful?

LOIC: What makes museums unique is the collection they have in their care, and the incredible body of scholarship they have developed about that collection. Those are our unique assets that we have to play with and to create value out of. This is abundantly clear with a physical experience of the collection at a museum – it is true also in the digital space.

In the digital space, there is more competition. We need to be very aware that there are people serving variants of our mission without a building, without a collection, or serving our mission in different ways. In order to successfully compete in these spaces, we need to amplify to our users that which makes us unique. Seen through this lens, I am cautious about how much of the storytelling a museum needs to do about its collection: it certainly does not need to be the "authority." I think there is a more impactful role for museums by being a platform on which global audiences can connect the collection and scholarship to ways that are most resonant to them. Especially in a world which is "post-truth," and where people are looking for genuine, factual content, we have that in museums.

KATI: For me, the biggest opportunity for us is creating experiences that matter. We are so well placed to create experiences around our collections and knowledge. Those can be physical experiences that are mediated through digital means, or they can be online experiences. Our advantage will be in truly understanding our audience and building experiences around their needs and interests. We're fortunate that we can do it at scale, and with such a curatorial depth of knowledge and expertise behind us – we are uniquely placed.

KEIR: And if we can help create the experiences that are positive and transformational. I saw Alejandro Iñárritu's *Carne Y Arena* at *LACMA* [*Los Angeles County Museum of Art*] last week, which is phenomenal. It's a three-part, immersive theatre performance about the act of migration from Mexico to the U.S. that uses VR. It's immersive and confronting, achieved in a way that I think is incredibly hard to do in another context, in that it is an artwork in a museum that you can fall into. If those are effective contributions to the digital future of museums then they, for me, hold immeasurable value.

KATI: Exactly. Exactly.

KEIR: Cool. This has been super fun. Thank you both so much.

One key aim of this book is to explore how practitioners can successfully work with or within the post-digital[7] museum (Parry 2013). This conversation surveys that territory from a particular vantage. The scale and resources of the *V&A* and *The Met* are large in comparison with most organisations in the global GLAM sector. However, there are numerous replicable elements in how Kati, Loic and their colleagues bring digital product development processes, ceremonies and language into the museum context. In particular, we draw your attention to their focus on the public-facing goals of open access, creating different modes of access and engagement, and valuing a purely digital interaction with their collections and content, as well as the internal process-oriented and organisational concepts.

Understanding and measuring digital impact is an active topic of discussion and research in the museum technology sector, with initiatives such as Europeana's *Impact Playbook*,[8] the 2015/2016 Visitor Motivation Survey of 18 museums across Australia, Canada and the U.S.[9], the *Cleveland Museum of Art*'s *Art Museum Digital Impact Evaluation Toolkit* (Bolander et al. 2018) and numerous other projects and publications. Kati and Loic explore how it's possible to create a sense of relevance, and even approachability, when working at a massive distributed scale.

---

7 The term "post-digital" museum has become more common in the discourse in the last five years. Ross Parry's article "The End of the Beginning" (Parry 2013) frames our thinking about on this term.
8 Merete Sanderhoff, who appears in Conversation 5, contributed to and utilises the playbook. Learn about and download the *Europeana Impact Playbook* at http://web.archive.org/web/20190331153323/**https://pro.europeana.eu/what-we-do/impact**
9 The National Museum Website Visitor Motivation Survey was conducted with 18 museums in 2015 and 2016, and presented at multiple conferences (Wambold et al. 2016) with more organisations joining the survey is subsequent years. All museums used the same questions and compared their responses following John Falk's Predictive Model for Museum Visitation.

For more on open access and digital production at scale, we suggest you read:

>Conversation 4: Sarah Brin + Adriel Luis
>Conversation 5: Sarah Kenderdine + Merete Sanderhoff
>Conversation 7: Shelley Bernstein + Seb Chan
>Conversation 11: Tony Butler + Lori Fogarty

For more on digitally informed processes and organisational structures, we suggest you read:

>Conversation 7: Shelley Bernstein + Seb Chan
>Conversation 9: Brad Dunn + Daryl Karp

References

Bolander, Elizabeth, Hannah Ridenour, and Claire Quimby. 2018. "Art Museum Digital Impact Evaluation Toolkit." Office of Research & Evaluation, Cleveland Museum of Art.

Cormier, Brendan. 2017. "How We Collected WeChat." V&A Blog. September. http://web.archive.org/web/20190620070332/**https://www.vam.ac.uk/blog/international-initiatives/how-we-collected-wechat**.

Parry, Ross. 2013. "The End of the Beginning: Normativity in the Postdigital Museum." *Museum Worlds* 1 (1): 24–39.

Price, Kati, and Dafydd James. 2018. "Structuring for Digital Success: A Global Survey of How Museums and Other Cultural Organisations Resource, Fund, and Structure their Digital Teams and Activity." Proceedings of Museums and the Web 2018. http://web.archive.org/web/20181019220314/**https://mw18.mwconf.org/paper/structuring-for-digital-success-a-global-survey-of-how-museums-and-other-cultural-organisations-resource-fund-and-structure-their-digital-teams-and-activity/**

Tallon, Loic. 2017. "Digital Is More Than a Department, It Is a Collective Responsibility." The Met Blog. October. https://web.archive.org/web/20190712194145/**https://www.metmuseum.org/blogs/now-at-the-met/2017/digital-future-at-the-met**.

V&A Research. 2018. "Immersive Dickens: Prototyping a Mixed Reality Immersive Display of a Charles Dickens Manuscript." V&A Research Projects. 2018. https://web.archive.org/web/20190712194305/**https://www.vam.ac.uk/research/projects/immersive-dickens**.

Wambold, Sarah, Marty Spellerberg. 2016. "Falk Meets Online Motivation: Results From a Nationwide Survey Project." Museums and Web. 2016. http://web.archive.org/web/20170804154016/**https://mw2016.museumsandtheweb.com/proposal/falk-meets-online-motivation-results-from-a-nationwide-survey-project/**.

CONVERSATION 7

Shelley Bernstein + Seb Chan

"We talk about 'time wallets' rather than cash wallets quite a bit. . . . It's actually *time* that prevents people from attending, and we see that across all audiences."
— *Seb Chan*

FIGURE 7.1

Shelley Bernstein (U.S.) spent almost 17 years at *Brooklyn Museum* where she spearheaded a series of innovative initiatives including: ASK Brooklyn Museum, an app that allows museums visitors to communicate in real-time with experts; *GO*, a community-curated open studio project; and *Click!*, a crowd-curated exhibition.[1] At *The Barnes* she instigated a series of internal digital projects, ranging from wearables to artificial intelligence, and helped develop public-facing partnerships with local organisations that brought their collection into the streets.

Before joining *ACMI* in 2015, Seb Chan (Australia) led the innovative digital program at the *Cooper Hewitt Design Museum* in New York during their 2014 physical renovation and digital renewal program, including the design, development and roll-out of the *Cooper Hewitt Pen*.[2] Previously, Seb spent more than a decade at the *Powerhouse Museum* in Sydney, Australia where he led pioneering open collection projects, enabling the *Powerhouse* to be the first museum in the world on Flickr Commons.[3]

Shelley Bernstein and Seb Chan have been active in the museum technology space for more than 10 years. They are visible members of this community through their publishing and speaking engagements, and for the highly publicised digital projects they've led. In late 2016, they engaged in an asynchronous discussion by commenting on each other's posts on the online writing platform Medium, having appeared at conferences together in the past. This short comment thread, with its open and generous back and forth, suggested a real-time conversation would enable a rich and nuanced discussion.

This conversation was recorded on May 18, 2018. At the time, Shelley Bernstein was Deputy Director of Audience Engagement and Chief Experience Officer at *The Barnes Foundation* in Philadelphia, U.S. and Seb Chan was Chief Experience Officer at the *Australian Centre for the Moving Image* (ACMI) in Melbourne, Australia.

KEIR: You are both Chief Experience Officers – a role that's become more common in the last few years. What role do you play in your organisation's strategic planning?

---

1 Details of these projects and the strategies and practices that enabled them, have been documented by Shelley for the *Brooklyn Museum*: https://web.archive.org/web/20190208070938/https://www.brooklynmuseum.org/community/blogosphere/author/shelley/
2 The Pen is a physical interface developed by and for the *Cooper Hewitt* that allows visitors to draw on digital touch tables and to select and save objects from the collection into a web-based record of your visit to the galleries. See: https://web.archive.org/web/20160328103020/https://www.cooperhewitt.org/new-experience/designing-pen/
3 The Commons on Flickr aims to display hidden treasures in the world's public photography archives and to enable public input into these collections. In this capture from 2008, you can see the *Powerhouse* and the *Brooklyn Museum* are two of the first museums on the platform. https://web.archive.org/web/20080805061656/https://www.flickr.com/commons/

SHELLEY: For us, it's about thinking about new audiences. *The Barnes* moved from the suburbs to central Philadelphia in 2012, and it went from 60,000 visitors a year to 260,000 almost overnight. But it retained the exact same audience, just more of them. I think Thom Collins, our Director, was interested in how to take audience-centric thinking, specifically about those who are not yet taking advantage of the institution, and embed that throughout the organisation. A lot of my role is questioning the things that we think we know and injecting the perspective of the audience, specifically a new audience, to challenge that.

My title is interesting. *The Barnes* was not ready to just go with Chief Experience Officer. They felt like a deputy director title was needed. . . . They wanted to integrate the position within the organisation and they felt like Chief Experience Officer didn't do that well enough on its own, so my title became Deputy Director of Audience Engagement and Chief Experience Officer.

KEIR: Seb, it's obviously very different for you.

SEB: Yeah, sure. My role is really a strategic experience design role across the *whole* organisation. It's as much focused on the way we work with our audiences and visitors in our communities outside our building, as within the building. And it's about the way staff work as well. There are elements of my role that you might find in HR/organisational design roles, regarding the tools and methodologies that staff use to communicate, project manage and collaborate with partners, coworkers, and consultants and exhibit design firms, particularly across multiple physical sites. There's lots of executive management stuff, and bits of architectural design, working with the exhibit teams. Lots of prototyping and piloting things with those exhibit teams, and large amounts of work with marketing and design. At *ACMI* we have a very good marketing team who work a lot with data and think a lot about content production and experience as part of marketing practice.

Then there is public-facing stuff, which is about engaging new audiences who have no experience with our building and spaces, and engaging current audiences in new ways. We have one and half million visitors a year, so we deal with some challenges at scale.

We are a building that is spread over four levels in the centre of Melbourne. One of the bigger challenges that we have is three different key audiences. Often the people who come to our free exhibitions don't go into our paid exhibitions, and people who come to our paid exhibitions don't realise we have massive free exhibitions. And neither of those audiences realise we have cinemas. We also have the cinema patrons who don't realise we have free exhibitions that are related to their cinema program. So, I'm really looking at ways to integrate those audiences and change their expectations and the way they experience the building.

We are not this type of organisation where we have subject matter curators. Instead, we have exhibit curators. They think about our audiences first, and they already think a lot about storytelling, so it's about more of a guiding role with them.

These days, other people are my hands, but I don't manage a huge team. I purposely have a very small team. We deploy like internal consultants – like a strategic consulting firm might approach a company and be flown into work with the team, change practices and move onto another team.

At the moment I am working on our renewal project. It's about 80 percent of my time, working alongside the chief curator, helping shape the stories and the curatorial work which all inform the resulting future experience. My role is a weird role that the name doesn't really fit anymore. We don't really have any other name for it, so it is what it is.

KEIR: Shelley, is *The Barnes* a hierarchical organisation? Are you able to work laterally as Seb is at the *Australian Centre for the Moving Image*?

SHELLEY: Yes, my position is at the top of the organisation chart, but there are a number of people at that level. I am often in everyone's cookie jar. It's an interesting place to be because you are often the person in the room going, "But wait, why are we doing it this way, why aren't we thinking about this, how do we do it more holistically?" It is not an easy position to be in. When I started nobody really knew what the role meant, and there was an adjustment period to figure that out.

KEIR: Like you, I am currently tasked with creating or fostering projects that reach and attract new audiences. However, the nature of many museum collections, their buildings and locations, their admission charges, mean that although they are open, they are not welcoming to many audiences, even in their local community. Despite that, most museum staff want their organisations to be open to all. Neal Benezra, Director at *SFMOMA*, often says, "Our job is to mean more, to more people."

Could you describe a project that sought to make your organisation more open, and more welcoming to those previously excluded or unengaged, and discuss what has, and hasn't worked about?

SHELLEY: *The Barnes* is the perfect example of that challenge. The median household income in Philadelphia is $34k a year, yet we have a $30 ticket price, which is the highest in the United States. You have to start thinking about who that keeps out inherently, right?

What you see is us designing programs to figure out how to programmatically make the institution more accessible in a targeted way. For example, we are doing a project called *Let's Connect Philly*, which challenges Philadelphia's artists to respond to works in *The Barnes*' collection by creating their own 8x10 piece that gets displayed and voted on. There is a prize – artists in the top four get residencies at *The Barnes* – but the program is designed to give an entry point that was not there before. Participation means responding to works in our collection and showing us your own view.

The project design had some very specific angles to it to target audiences we wanted to grow relationships with. Artists could only participate if they lived with the Philadelphia County line. Immediately we started to engage people inside the city who we were not engaging before. It gave those artists free visitation over a five-week period. We saw 517 artists visit. You start doing that math and you realise that is close to $20k in admission. By targeting it, we are saying, this is an audience we want to cultivate. Then, how can we be accessible to them?

The next stage of this project invites anyone to come in and vote, and when they do, they get free admission to *The Barnes*. It's a way to activate those artist networks already within the city line and engage a local audience that might not normally be coming.

I believe that price is a real barrier. I don't think every institution has to be free, but we do need accessible pricing models. Having worked in NYC for so many years with many museums at "suggested admission," I now have a much greater understanding of that structure's ability to grow audiences in ways that don't just rely on value arguments. Until we have accessible pricing models, the issue will become that we have grown an audience for single programs that can't easily convert into the general admission or membership, and that's a huge challenge.

KEIR: Thanks, Shelley. I think the *Let's Connect* project also does something, albeit quietly, which is address the overwhelmingly white male disparity of the canon.

SHELLEY: Yes, I agree, because as an artist you can take what is on our walls and make it your own. So many of the submissions had the power to take a collection dominated by Western art and Western voices and counter it with much greater diversity. And that diversity is what makes Philadelphia great. The project really combined the two in a very modern way, driven by the voices of our local communities.

KEIR: Seb, could you describe a project you've worked on that is about welcoming new audiences?

SEB: Yeah, sure. Look, it was fascinating working in America and seeing how difficult America is, and then coming back to Australia, and working with museums in other parts of the world, too. I have to disagree that price is a major deterrent. It is time and interest. Free time is a very significant marker of economic status and class. If your family needs to work two jobs or more, you are not going to have time.

We talk about "time wallets" rather than cash wallets quite a bit, because the time wallet and how you decide to spend that time wallet is determined by what other things you have to do. Family you need to care for, community groups you are engaged with, for example. It's actually *time* that prevents people from attending, and we see that across all audiences.

Most of the activities that my museum does now are actually free. We have free permanent galleries and those galleries are open from 10:00am to 5:00pm. Our cinemas charge, but at a very low cost compared to commercial cinemas. Those cinemas run from 10:00am until the last session that goes in at 9:00pm. We have talked about ways of running the museum to synchronise with the cinema hours, but currently we just can't afford to.

Our main revenue source is government, so we are tied to – and in support of – government policies. In [the state of] Victoria we have centre-left progressive state and city governments. Even our centre-right federal government would be a much more Democrat government in terms of American political spectrum. It's a different context. That's not really answering the question, but hopefully problematizing around price and accessibility.

The other thing is we are not a museum that is about static collection or even about a collection necessarily. Our permanent collection is not in our permanent galleries. We are a museum that covers a series of topics and how those change. We show lots of different works going back in time, but also very contemporary works. We commission lots of works. We just finished judging the second virtual-reality [Mordant Family] commission series. The first was by Indigenous artist Christian Thompson and the second is done by a woman named Joan Ross, who does post-colonial work, and she will be making her first VR work.

We host a lot of film festivals. We've got the Human Rights Film Festival on at the moment. Queer [Film Festival] was on a month or so ago. So, we have lots of different, diverse communities coming through our building. What we are trying do is get them to engage with different *parts* of our program. That interconnectedness of programming is one of the bigger challenges that we are working on.

Of course, we also have big access challenges like everybody. We had a TV series in Australia called *Sunshine* which was about Sudanese youth in Melbourne, it's a youth social commentary drama theme. I guess like . . . what would you call *The Wire*? It was that sort of drama.

KEIR: Gritty social drama?

SHELLEY: Drama.

SEB: Social drama, anyway. So, Sunshine is a suburb in Western Melbourne, with lots of new, migrant communities. Sudanese youth are getting a lot of terrible mainstream media coverage at the moment. They are targeted by the police, et cetera. This TV series is produced "in community, by community'" with a largely non-professional Sudanese cast. We did the premiere screening – directors talked and the Sudanese cast talked. Really great, and it was the first time some of them had even been into the city centre, let alone our museum.

The challenge is getting them back, and it's not just about content, it's about fitting into their lives. It comes back to time, and preferences that people have of how they spend their time. That value, that social currency – [French sociologist Pierre] Bourdieu's notion of "distinction" and social capital.

KEIR: I often have disagreements with my colleagues who look at other museums or cultural organisations and say, "What are they doing to attract the audience we are trying to attract?" And I say, "No, no, it's Netflix and sunny weather that we are in competition with." And that is for people who have time to spare. People who don't have money in their time wallet, who are working two jobs, are parenting, are traveling or commuting, or shopping. Doing all the things that don't leave room for us . . .

SEB: Going to church.

KEIR: Going to church, exactly. I think there is a grave misunderstanding within the museum sector that there is some sort of model cultural visitor who wakes up and thinks, "Which of the cultural institutions in this fair city will I go to today?"

SEB: We talk a lot about sport here in Australia.

SHELLEY: Yeah.

SEB: The AFL (Australian Football League) in Australia has done a huge amount of work with First Nations communities, and now Indigenous communities are very much a part of the public face of the AFL. And in the last few years, the women's AFL has been blowing up here. And it's not been a diversity program, it's just been about keeping the sport alive. They have been massively successful. There are still huge problems with race and gender issues, all sorts of things, within the AFL, but there is a lot that can be learned from how they have done that. Genuinely grassroots, and all the way to the top.

I think that has a lot of parallels with the work you have done, Shelley, with artists. It's almost like a "grassroots training program for artists." It's interesting to look at sports, and how sports have dealt with this.

KEIR: Both of you work "in the open," so you talk and write a lot, often about things that aren't finished or are a work in-progress. What is the strategy behind that approach? What do you get out of it? What is it doing for your organisation or your professional practice to be transparent about your process and outcomes?

SEB: For me, it's about working in public. I grew up with the web, and this notion of "viewing source code" has infused my philosophy around lots of things. That we all get better if we share information, and we bring the other generations along with us. And it is about transparency and openness as a differentiator, too. I found that early on at *Powerhouse*. It was a way of creating an international network of peers and peer institutions that we could bounce ideas off and share information with.

In the early days, that often created issues with marketing departments who want to "control the message." There was debate around who speaks for what. A discomfort around the outsized personalities that some of us became.

I have tried, with the institutions that I have worked with, to make [working in public] part of what they do. At *ACMI*, we have the education staff, the philanthropy staff, everybody trying to do this too. In my mind, the institutional brand wherever you are should be made up of the voices of all of the staff, who all have personalities beyond the institution. I think that it is important that that we are all brand ambassadors, or whatever the horrible term is now – particularly for us in Australia and New Zealand. Workers in Australia and New Zealand are so far away from the rest of the world. There is a lot of amazing stuff that happens down here, often ahead of America and Europe, and we don't talk about it very much.

There are a bunch of Australian and New Zealand-born museum technologists – like yourself, Keir, Tim Sherratt, Fiona Romeo, George Oates, Courtney Johnston, Andy Neale, Fiona Fieldsend, Mia Ridge and Mitchell Whitelaw – who have become known globally in this sector because of talking about things, or trying to bring others along with us. That's put Australia and New Zealand on the digital map in the cultural sector. And also set things up for the next few generations for whom digital (and technology) isn't special, it just is what is.

KEIR: Yeah.

SEB: That's a really positive thing and it changes the way the organisation feels. In my career, it's brought lots of other people from all different fields into the sector as collaborators, from medicine and health, from computer science, all these other realms. I think that's been part of changing how museums think about themselves and how they might think of it as a business strategy – competing by collaborating. And I think the "next generation" piece is absolutely critical – mentoring and working with students and emerging professionals, bringing them in and making their voices heard amongst the cacophony of other stuff.

KEIR: We have a policy within my department that anyone who works, even as an intern, will get published no matter what. I think that's critical. We pay our interns for their time, but you can pay in many ways, and visibility is one mechanism.

SEB: Totally. Across the whole museum we have a professional development plan that goes into re-grades and bonuses. Part of everyone's professional development plan is writing or talking about their projects, and their role in projects, and building those writing and communication skills across the whole organisation. The biggest push back from managers has been around time commitments, but the project isn't finished until staff have written about it, at least internally. And that's not my staff, that's staff across the museum.

Using blogging or writing as a reflective practice is important, and something that I encourage everyone to do.

KEIR: Shelley?

SHELLEY: For me, it's about processing and reflection. You will see me talking about times when I don't know the answers or times when we absolutely failed. I think readers have been interested in that transparency and openness. To have somebody say, "Wow, we just hit a wall here, and here is how we are trying to think about it, and work around it."

I'm always amazed at what gets read and what doesn't – this has taught me a great deal. Some of my posts don't get read at all, and then sometimes I'll put something out that gets read like crazy. The most recent example of that was a post about challenges on our front lines with people getting too close to artwork and how we interface with them. We adopted "authority of the resource," (Bernstein 2018) which has been around forever, so I was like, "Well, this is the dumbest thing for me to post about because this is old news." But also, "You know what I have been struggling with this for a year. It's time to write this because we have gone through so many iterations and I might as well just put out in the world." Everybody is talking about it like it's new material. And I am like, "Wow, everybody must be struggling with this."

KEIR: Yep, we all are.

SHELLEY: I didn't know this thing that I had been dealing with in my little microcosm over here in Philadelphia was something that everybody is dealing with until I put it out there, and saw people talking about it so quickly. To see what hits a nerve is helpful for me as a professional.

KEIR: Sometimes this dialogue is happening in places that you are not aware and may only come back to you the long way around.

Changing topics. As you know, I am in San Francisco, home to utopian ideas about technology, but also home to platform makers like YouTube, Google, Facebook, Twitter. They build on the open standards of the web, but I consider them closed systems. Something I have been thinking a lot about is how our organisations pour information – even their collections – into them, only to see them disappeared, abused or deprioritised behind the platform's own content, in the case of say the Google Art Project or the Google Cultural Initiative.

With that in mind, what do you think through when you are contemplating adopting technology platforms for your organisation?

SHELLEY: Seb, you take that first.

SEB: The reality is that most of the open-source tools used in science, data sciences, digital humanities and other fields that are used by lots of people are maintained by teams smaller than three people. Massive, fundamental open-source scientific measurement tools are maintained by teams smaller than a museum's technology team, and used to control satellites, do cancer research.

Shelley and I were both very early adopters of Flickr and Flickr Commons, and it came down to a handful of people. It will always be a handful of people who really care and have it made part of their work. I think we need to be honest that the business models that support many things that we do are not the same as the business models that support most commercial enterprises and they are very, very fragile.

In this late-stage capitalism, everything is a Ponzi scheme, everything is fragile and building resilient systems comes down to resilient people. This is why I am always concerned about the departure of good people from our field, because when the good people from our field go, lots of other things go that perhaps we haven't realised. Regarding [other people's or commercial] platforms – we have seen them as more stable and more resilient than they really are, and I think we are seeing this with Facebook now. Facebook's resilience is not actually very resilient at all. Similarly, Google's business models are super fragile. They are just really big [in scale].

The capital investment needed to do good things within our sector is very difficult to get for technology initiatives. The work coming out of Australia, New Zealand and Europe, where stability of funding comes from taxes and government support, is where you want to look to build alternative platforms. Then use platforms with less stable business models, ones based on the sale of user data, which have a short-term life compared to museums and libraries, as a secondary place that you'll do things in.

But by the same token there is a lot of stuff that we do that doesn't need to last forever. The three of us [also] do marketing initiatives. If we were working in any other field, you might call them "brand activations." But the stuff that really matters, the stuff that we want to last for a long time, we should be building our own platforms within our fields and building stability around them by developing with the right people and building them into the core business model, or the operating model of the institution.

What has been interesting at the *Cooper Hewitt* is that the systems we built we built very well, so even though all the technical staff left, it is still running. It has been running basically on minimal technical support for nearly three years. I would count that as a success. But the reason that it's still running and the reason the curators are still putting all of the loaned pieces into the system is because it was well designed – it became part of the operating system of the museum.

So, it wasn't fragile, it didn't depend on me, or Aaron [Cope], or Micah [Walter], or Sam [Brenner] or Katie [Shelly] being around. It was fundamentally built into the operating model of how the museum, how curators, should work. So, you could remove the entire digital team from the museum for an extended period of time and it would still work. That is a system that isn't fragile. It doesn't, however, mean it's indestructible.

KEIR: I want to respond to the point about fragility. Earlier today I deleted the *SFMOMA* app from the App Store because Detour was bought by Bose, and Bose is turning off all the services by the end of May [2018]. We have a big Rene Magritte exhibition opening today, so we timed our replacement of the Detour app platform to the opening of the Magritte show. We have deleted the old app and the new app is live. We had about five weeks to go from the old app platform to the new app platform, and now we are returning a very old school idea . . . you punch in a number and it plays content.

SHELLEY: Oh my god.

KEIR: It's an interesting moment of fragility. We won all the awards. We served millions of minutes of content to visitors, had hundreds of thousands of downloads. It really worked, and it's gone. It was literally, "Are you sure you want to remove this from the App Store? Click here to do it." Click . . . done. So, I take your point about fragility, and that's with the team still in place. It was simply out of our hands.

SEB: It was a great brand activation!
   The Pen was a great brand piece, and it was built into the rebrand of the institution and the repositioning of *Cooper Hewitt*, and it was built to be non-fragile. But others in the field don't see that. Senior leadership often just see the brand piece. Others just see the tech piece. It's the integration of those that makes things stick around.

KEIR: Shelley, how do you pick your platforms? What do you think about before you take on platforms? You have done a lot of experimentation in the space.

SHELLEY: It's funny to bring up Flickr again, but Flickr made things easy in a couple of ways. It was fragile because of its business model. But they gave you a way to do bulk uploads, right?

KEIR: Yep.

SHELLEY: Through a non-technical means. I am not a coder and I uploaded all of our Commons images. And at the same time, when it was time to leave the Commons, they had an API that let us suck down every bit of information and put it in our collections database. When I look at anything developed outside of us, I am going to assume it's fragile and then, based on my timeline, I am going to say, "How hard is it to do it?" For instance, in the early days of the Google Cultural Institute, it was a painful to upload anything.

KEIR: Really bad.

SHELLEY: And I will say the same thing about iTunes. Remember iTunes U and how crazy that was to just upload content? In the end we pulled out of iTunes. I was like, "Forget it, this is crazy, this is taking us too much admin time." When I moved to *The Barnes* there were a number of reasons why we didn't go into the Google Cultural Institute, but the only thing I cared about was street view, because it was the only thing that would get us on that platform that didn't require an insane amount of admin time to do the care and feeding. And never mind when your collection records changed – there was no way to get information changed in that platform.

SEB: Yeah, of course.

SHELLEY: I look at that stuff and I am like, "Is there an API? Can I get my stuff in and out? Can I get changed records in and out? And, can I get information in easily if I am not a coder?" All of that changes how we think about the platforms. For instance, I have a short-term exhibition, I have to do a microsite, god help me I am not going to use anything that isn't Square Space, Wix or similar. The assumption is it will be around for a year and then we say goodbye to it. What it the easiest platform to help me do that?
   Third-party platforms have been incredibly helpful to somebody like me who doesn't have a specifically digital staff. For example, we are using something called Formstack to control all of the workflow for *Let's Connect*. Artists submitted work using forms we built on that platform and it was amazing. 800 artists coming in the door in a 48-hour period and we could look up their submission, check them in, have them electronically sign drop-off permission, and send the whole record back to them by email. No

coding. And all I need is for Formstack to work for another two months and then *Let's Connect* is over, and I will extract all that data and not care about the actual platform.

So, in the end, I look at where we need long-term architecture and where we don't. When we don't, I look to third-party and, specifically, which third parties let me get data in and out easily assuming the platform is eventually disposable.

KEIR: Weirdly, the talk that I gave when I first came to the U.S. seven years ago about some work I did at the *MCA* [Museum of Contemporary Art in Sydney] was about simple systems, best of breed, loosely coupled, without a big team.

SHELLEY: Yeah.

KEIR: There is a sort of fetishism around the shiny new thing, and often it's about being thoughtful about how you link the chain up, what you are using the chain for.

SHELLEY: I was lucky that I had a development team of about seven at Brooklyn, including Paul Beaudoin who developed the collection online. One of his first things was to build the API that we then used for everything. That was just one developer saying, "No, no, this is how we need to do this." And it made that practice sustainable across so many projects.

I believe when we think about time and investment, investing in that thing internally that is going to then get you to many places with less friction is the best place to put the effort.

KEIR: Agreed.

SEB: The other part of my work at the moment is around digital preservation. My museum collects film, TV shows, so [almost] everything is in copyright and everything post-2000 is digital from the start. We have got Soda Jerk's *Terror Nullius* on at the moment – one of our commissions, if it's ever in your city you should see it, it's awesome. . . . In the past, we would acquire a copy of the film and preserve the film. Nowadays, we want to collect the digital file. But I am [also] interested in collecting the Final Cut or Premiere project. It's appropriation art, a compilation of 300–400 different Australia and New Zealand films that speak to each other. Characters are cut into each other. It's composited together. For us, preserving the artwork also probably means preserving a copy of [Adobe] Premiere that can run it. The object is now part of a system and we're preserving that system.

I think museums take their, "we have to keep it forever" mentality and it dribbles down into all the projects we do. If we are honest, not all the things have to live forever. Not all of the things should, but there are some that should.

We may need as a sector to actually decide that we need to collectively own and make things together. It seems to have worked okay in Europe and certainly if we look in Australia, we look at Trove, which is a meta catalogue that the National Library built, and Digital New Zealand, those sorts of things may be where there needs to be collective investment. I don't know how that works in the U.S., with its complicated governance model and states.

I mean the *Smithsonian*, too. They could be doing a lot more impactful work than they are now. Collection management is a problem we all have, all the collection management systems suck. Why isn't this problem being solved by people in the field?

KEIR: Yeah, but it's doable.

I think everyone in the Western world is now more aware of data and privacy. This is after the breaches at Equifax, Target and Yahoo.[4] When will the first museum CRM be hacked, and what will the ramifications of that be?

SEB: They have already been.

KEIR: Oh yeah, really?

SEB: I was part of a panel at AAM [in May 2018] and there was a question in the panel, "How many of you have had a data breach, significant, minor, don't know, or none." And there were a couple that put their hands and said they had major data breaches, so it has happened in our field.

KEIR: This hasn't come out publicly, other than hands up in a conference.

SEB: Yeah, because it's like, you know, the number of people who actually visit museums is actually tiny. So-

SHELLEY: Tiny.

SEB: The database is like 50 people, or 50,000?

KEIR: Okay, so in that panel, what were you trying to communicate to people in the museum field about data and privacy? What should they be considering? What are they not considering? Where are our blind spots?

SEB: Collect as little as humanly possible, store even less, don't ever require people to trade their personal data for access to things, make that an add-on after you have given them what they wanted.

You wouldn't believe the debates we had at *Cooper Hewitt*. My team was very much, "give everybody their things and if they want to create an account let them do that." Many people from all over, outside and in, wanted to make web visitors register first, claiming that otherwise they never will. "Why do we want them to register?" I'd ask. "Well, we want to do stuff with them." Actually, staff [asking for registrations] just want a big database, and it becomes like nuclear waste. Visitors and people *do* register when we give them value first, and then they decided they wanted to be part of the system.

---

4 Making news at the time of this discussion were the Equifax data breach where the personal details of 147 million Americans was stolen, the revelation and subsequent financial settlement from Target's 2013 data breach where its database of customer's personal information was hacked, and the latest in a series of user account information breaches at Yahoo that totaled in the hundreds of millions of accounts.

Collect as little as possible, is my advice. We have mandatory data-breach reporting in Australia which you have got to do within 48 hours. The GDPR [General Data Protection Regulation] will be everywhere soon.[5]

We have an experience currently in one of our exhibitions [at ACMI] where people scan their map and the map is magic and it makes a character. Visitors can scan their face and their face gets put on that character and it does stuff. One of the developers involved was like, "Oh well, you can do some good data analysis of who comes to your museum. We have all their faces, what do want to do? You could run computer vision across the image set." Of course, I have thought about this. "Like, seriously you could do gender, race, emotion, all that stuff. It would really great, and its kind of anonymous because it's only their face, we don't know their name." I am like, "Bin it, bin all the data as soon as it's off that wall." We don't need it. The *visitor* needs to see it on the wall. Take their selfie with it, then send that photo home. We do not want to keep that photo. Yes, there's all these ways of extracting value from it, but the customer hasn't opted in, we haven't made that part of the deal. If they have opted into that maybe that's then fine, but they haven't.

Be very, very explicit. With machine learning and computer vision, people are just doing things because they can and it's going to be pretty bad actually.

Take as little as possible and be upfront about what you're doing with it.

The way the *Exploratorium* has done this has been really great. They show what they do with the data. When you signed up for things on their website, or when it collected cookies, part of their museum teaching was showing what happens to data. They are a science centre so it was part of their mission and purpose. *Big Bang Data*, an exhibition that the Barbican did, had similar things.

There is an opportunity for museums not only to behave well, but also to teach people about surveillance and data collection, no matter what sort of museum you are. You don't need to be a science museum to talk about surveillance. You can have a historical collection and talk about surveillance as well. It's one of the key issues we should all be talking about, and as a progressive field it should be one of our number-one topics.

KEIR: I don't feel like it's being talked about at all. Other than in meeting rooms where people are arguing about UX, when actually it's a fundamental decision about how you treat your audience. There's a "checkout as a guest" on our website, and I had to make a data-driven "this will make us more money" argument in order to allow us to have anonymous, non-capture checkout.

---

5 Arriving in May 2018, the General Data Protection Regulation is a set of data protection rules for all companies operating in the EU and used by European citizens, wherever the company (or museum) is based. It includes much stronger rules on data protection and user control over personal data that previously, or currently exists outside Europe. The introduction of the GDPR affected organisations across the world.

SEB: It's more of an issue in Europe, I think. To be honest, with the GDPR, everybody is talking about it in the U.K. In Australia, people have been talking about ever since I came back. It may just be in the U.S. that it's not been.

SHELLEY: I will say I love the internal discussion happening because of GDPR. We have a really solid lawyer who is looking at data protection and adoption right now. It's the only time that I have been able to win over the marketing effort. If you are in a revenue-minded institution, then they want to keep all the data because we can possibly monetise it. It's the lawyer who is coming in and saying, "No, no longer." That is fundamentally changing a dialogue I thought I would never have the power to change.

SEB: Yeah.

SHELLEY: For the longest time when visitors stepped into *The Barnes*, the staff would ask people hand over their email address in order to buy tickets. There was zero transparency around it and people would just give the email thinking it was a necessary part of the transaction.

KEIR: Whoa.

SHELLEY: The staff wouldn't say, "Hey, would you like to join you to our email list?" They would say, "What's your email address?" Then into the system it goes, and you've joined a mailing list without opting in. I don't think that's responsible to our visitors and, sometimes, it helps when you've got backing to do the right thing even if it is as a result of legal requirements.

KEIR: A lot of it is about making decisions that protect against an unknown future. With *Send Me SFMOMA*, we received text messages from hundreds of thousands of Americans expressing a desire to engage with art via SMS. The code anonymises every phone number on receipt, makes a hash of it and only stores this wacko hash. So, five years from now, no one can text all those people and ask for a donation.

SHELLEY: During *Go!*, which was that open studio project that we did in Brooklyn with 1700 artists, 36000 voters participating, we baked into the terms of service that we would not share any of their data with the *Brooklyn Museum*'s mailing list or any of the museum's partners. And everybody came to me internally to say, "Can I have all that data?" And I was like, "No you can't, actually we destroyed it, this is what we said in our terms." People were upset, but it protected us, and it protected the artists. And that's what you have to do.

KEIR: Maybe there's a difference between a brand activation, Seb, and the projects we do?

What norms within cultural institutions would you like to see disappear in the next 10 years?

SEB: That's a broad question. For me, it's about the divisions between the types of museums. I have now worked in all different types of museums. They are all the same to me, and I would go further and say archives and libraries are merging into museums. We have been seeing this trend in other parts of the world. The only thing that hasn't changed is philanthropic models behind them. Museums of art are tied to social capital and social status. That's what makes them different, not what they actually do in the world.

That outside view that a history museum is different from a museum of art, it's different from a science, it's different from a natural history museum, affects the audiences who come, and the way audiences think can and should behave with us. But it also affects the staff, and it affects the way the staff don't share across the fields. There is a lot of practices that large museum curators could learn from librarians, but they would never think to ask.

KEIR: Hmm.

SEB: There are digital preservation practices in the archives world, the art world, the science world, medical data preservation . . . if we come together, there are practices we could learn from each other that could move the field forward a lot faster than it is now.

And that is one of the fascinating things when I go to CAA (College Art Association) compared to AAM (American Alliance of Museums). I wonder, "can you all just get in the same room, because actually we are talking about exactly the same thing." The person in that library in Seattle has actually solved the problems of this Dutch social history museum, which has solved a problem for this Japanese science centre.

SHELLEY: It's not just how we view ourselves. It's also that the public doesn't view those distinctions a lot of time, and there are collaborations that we could and should be doing that bridge those gaps. For instance, at *The Barnes* we are doing a collaboration with the Bike Share, we are doing a collaboration with the Free Library where we are taking VR headsets of the collection into neighbourhood libraries and sending live teaching teams. We are doing a collaboration with Parks and Rec where we are sending our summer programs into the Rec [recreational] centers in Philadelphia, rather than expecting everybody to land on the Parkway.

You start to ask, "who is the person or the company in the city that's doing more interesting work around this specific thing than we are? How do we partner up to think about those strengths and learn from them?" The Indego Bike Share thing was an unbelievable opportunity. You had an organisation in Philly that had thought about diversity and inclusion from the ground up, and then you had *The Barnes*, which is a primarily white staff of a certain status. Just in trying to collaborate, the Indego staff would say things like, "Okay, we are going to crowd source the bike racks." And I'd say, "Okay, great." I would go to my staff and say, "What are the public domain works that we can put on the bikes?" And we would hand works to Indego, and Indego would turn around and say, "Okay, no, these don't speak to diverse audiences, try again." Just having that catalyst, that dialogue really changed our work for the better.

KEIR: Yeah, Dominic Willsdon and Deena Chillabi in our Education and Public Practice have brought a non-lending branch of the San Francisco Public Library and inserted it on the second floor of *SFMOMA*.

SHELLEY: Oh yeah.

KEIR: It's fantastic. They are doing library programming in the museum and commissioning artists to do artistic programs in the libraries. It's the most progressive thing that's happening here, but it's literally stuck behind the elevators.

SEB: We create a lot of barriers between our sector and other sectors.

The other thing I'd change is actually do marketing properly and not see it as the enemy. There's lots of stuff that we could be doing around sensible marketing.

KEIR: No comment.

SEB: The other piece, of course, is around paid salaries. There are big inequities in our field, and if we are going to get more diverse workers, we've got to deal with pay in our field, which may require us to also strike different partnerships. One of the things I was thinking about when I was at *Cooper Hewitt* was that the people who fund museums in the U.S. rarely attend museums in the U.S., and the people who attend museums in the U.S. are never part of the way the museum makes money, and pays their staff.

The philanthropists and donors who give money to the museum are not the average user of the museum. They are not the person you are designing exhibitions for, labels for, the education programs for. They might attend the opening, the gala, but they are not the primary user of the museums. But the user of the museum has no input into how the museum is funded. They have no financial levers of support. So maybe we need to make a new social contract with our communities, where the communities are able to directly support us, and we directly support them.

In some way the taxation model that we have in Australia and in Europe has an easier way of doing that. It's not the best way, I know – the strings that are attached with changing government policies. But that social contract piece is super important, and there is no quick solve to that. If the finances are coming from a handful of people, they have much larger sway over what you do, and the way it is done, than they should.

KEIR: To return an earlier point about fragility and sustainability. Sustaining the skilled staff that allow our systems to be successful long-term is a question of pay and equity. In the U.S., we often pay our junior staff lower than living wage. We pay them not much more than they would get working in fast food, but we expect them to have at least a graduate degree. There's no way to start from the ground up without some other sort of support. You end up only staffing museums with people who can afford to live on a museum salary. That's a fundamental problem.

SEB: One of the fascinating things in the Nesta report (Armstrong et al. 2018) that I contributed to was the calling out explicitly of spousal support and that arts diversity in the U.K. should not be left to spousal support or family inheritance. The way to get more diverse workers was to find ways to pay proper living wages, so you didn't require a spouse who was a lawyer or a banker to be able to even study in the field.

KEIR: Looking out beyond 2020, what do you think the dominant models of experiencing a museum will be? And will they be the same as they are now?

SEB: The dominant models?

KEIR: Yeah.

SEB: The *dominant* models will be exactly the same as they are now, because they are pretty much the same as they were 40 or 50 years ago. Hopefully we will have more different communities making use of them, and making more and different types of demands, and we will have more flexibility within how we work and are structured to respond to those demands. That would be great.

Fundamentally, I don't think it's actually going to change. If we do our job extremely well then different people will be added to the mix of users, and that will change how people encounter and experience things, art, ephemera, knowledge, all that stuff. But it also depends what happens with immigration, climate change, social disparities, whether there's cultural revolutions or social revolutions in previously stable nations. Those are probably the bigger issue, and whether a museum becomes a refuge in the way libraries have become.

SHELLEY: I think we should leave it at Seb's words.

KEIR: Me too, thank you both.

As Courtney Johnston states in her foreword, this is "a book that brings into conversation a group of thinkers who characteristically do their thinking in public." Shelley and Seb have both been leading proponents of this kind of practice, working through ideas in public and sharing their successes, failures and processes. In this conversation, we gain insight into this approach, which draws upon a kind of "view the source code" ethos of the early web, and gives countless others the tools, language and approach to bring the same ideas into their own institutions. This approach, which we subscribe to as well, serves as the foundation for reflective practice. Through these kinds of reflective techniques, students, theorists, practitioners and vendors can better develop, clarify and articulate their own thinking about digital technology in museums. It can also be an important tactic for institutions seeking to use technology to reinvigorate their brand, and to show relevance to contemporary and tech-savvy audiences.

Shelley and Seb also make visible that digital practice in museums today informs organisational behaviour across the museum, from providing tools and methodologies that enable internal and external communication and collaborations, supporting collections management and visitor engagement work. One exemplary practice detailed in this conversation is collaborations that push outside the sector, such as Indego Bike Share, Flickr and public libraries in San

Francisco and Philadelphia, and the freedoms and challenges that these sorts of collaborations afford.

For more on the link between digital practice and organisational behaviour in museums, we suggest you read:

> **Conversation 5: Sarah Kenderdine + Merete Sanderhoff**
> **Conversation 6: Kati Price + Loic Tallon**
> **Conversation 11: Tony Butler + Lori Fogarty**

For different modes and models of collaboration, we suggest you read:

> **Conversation 2: LaToya Devezin + Barbara Makuati-Afitu**
> **Conversation 5: Sarah Kenderdine + Merete Sanderhoff**

References

Armstrong, Harry, Hasan Bakhshi, John Davies, Georgia Ward Dyer, Paul Gerhardt, Celia Hannon, Svetlana Karadimova, Sam Mitchell, and Francesca Sanderson. 2018. "Experimental Culture A Horizon Scan Commissioned by Arts Council England." www.artscouncil.org.uk.

Bernstein, Shelley. 2018. "Using Authority of the Resource Technique at the Barnes." Medium.com. http://web.archive.org/web/20180912172131/**https://medium.com/barnes-foundation/using-authority-of-the-resource-technique-at-the-barnes-e6dca41f1b62**

CONVERSATION 8

Kate Livingston + Andrew McIntyre

"We've got more data than we've ever had, but we will still don't understand the audience. What we lack is insight."

– *Andrew McIntyre*

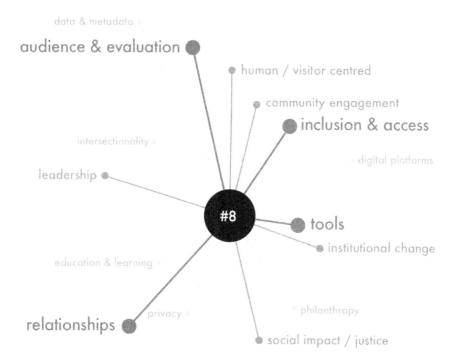

FIGURE 8.1

Kate Livingston (U.S.) and Andrew McIntyre (U.K.) each founded, and continue to run, businesses that focus on audience insights and evaluating museum experiences. Kate's practice is based in the U.S., while Andrew is based in the U.K. Morris Hargreaves McIntyre (MHM) has clients across the English-speaking world. We brought Kate and Andrew together as they are two of the most thoughtful and nuanced practitioners we'd encountered working in this space.

Both Kate and Andrew bring professional experience in the museum sector to the consultancy work. Before creating her own consultancy, Kate was the Director of Audience Insights at the *Denver Museum of Nature & Science* in Colorado, U.S. And before co-founding MHM, Andrew was Head of Research at *Arts About Manchester*.

Keir had been lucky enough to work with both directly. Andrew's company, MHM, helped *SFMOMA* evaluate its main institutional website (www.sfmoma.org), a project that Keir led, highlighting where that project had and had not been successful. MHM was also engaged by other departments at the museum. Kate evaluated three different interpretive galleries at *SFMOMA*, and her process encouraged Keir to think differently about visitor motivation, providing tools and strategies that he still employs. Keir's professional familiarity with their respective approaches helped enrich the discussion, which includes a mix of broad-ranging theoretical and highly detailed methodological moments.

This conversation was recorded on July 6, 2017. At the time, Kate Livingston was Principal and founder of ExposeYourMuseum LLC in Denver, U.S. and Andrew McIntyre was Director and co-founder of Morris Hargreaves McIntyre in Manchester, U.K.

KEIR: Andrew, what tools are most commonly used by Morris Hargreaves McIntyre (MHM) when trying to understand a museum's audience?

ANDREW: My journey into audience data started by connecting rather dull data, such as behavioural data, profiling data and demographic data. I realised quickly that such data is good for monitoring, but not for planning.

We started to develop areas of study that gathered information on why people are doing things. What are their motivations? What are the outcomes? What are the inputs and outputs of engagement?

That led us into the area of belief systems, values and everything that frames how people engage. We developed tools to explain why people do what they do, whether behavioural or looking at outcomes. The ultimate level of this is segmentation, which is an attempt to map the landscape of the audience so an organisation can understand how to engage people on their own terms.

Our culture segments are based on deep-seated cultural values and beliefs. It is hugely correlated to motivation and behaviour. And it is insightful in both designing experiences to appeal to those segments and in designing communications that motivate those segments to engage. That's what we've spent most of the last decade working towards.

It's a system that's being adopted in lots of different countries by different institutions. We try to turn data into insights, and then insights into models or categories, and then we try to turn those models into tools.

I think we're toolmakers.

KEIR: Kate, what are some of the tools and approaches that you find best deployed?

KATE: Something that Andrew said really resonated with me. I remember coming into the arts and cultural field and assuming that the tools, instruments and techniques would be fascinating because the setting is so visually rich, stimulating and engaging. I was dismayed to find how often it was rating scales and surveys, and other old school research and evaluation instruments.

We all want to better understand our audiences – their motivation, their engagement and other concepts, such as affinity or belonging. But when you talk to visitors about those concepts, they would never use a rating scale or checkbox. If I ask you to tell me about your feelings of belonging in the last museum visit you had, you would never say, "Kate, on a scale of zero to five. . . I would give it about a four."

There so many things that we can do beyond the basic survey, utilising technology tools like photography, videos, rich storytelling and narratives. It's not to leave out segmentation, because it is valuable as well. I would love to see more creative methodologies, but most of my clients first expect that we're going to do a survey. I have to convince them otherwise.

KEIR: Since I joined the museum sector, I've seen the focus moving from quantitative data and overall audience demographics to more integrated qualitative and quantitative data, including a shift towards psychographic segmentation. Do you think that's an accurate reflection? Is this limited to museums with big pockets or is it across the board in museums, and also libraries, aquariums or history centres?

ANDREW: The use of psychographic research is worldwide. Although it's not necessarily being adopted at the same speed.

KEIR: Right.

ANDREW: The biggest thing staff say is, "we've got more data than we've ever had, but we will still don't understand the audience. What we lack is insight."

There's a couple of trends that are sideways or backwards steps. As Kate said, the old-fashioned survey dies hard. A lot of people are diving headlong into big datasets, but what they have is a lot of big boring data. We've started talking about "deep data" rather than big data to explore how to start working with it. The magic is actually the humanisation of big data. It's the grounding in human experience.

Although there are lots of technologies you can use in creative research, interpersonal engagement and experience – actually talking to people – can expose decision-makers to people who are struggling to engage. I love the idea that we're going back to humanity. Realising that science and mathematics don't give us what we actually need.

There's another trend that is slightly disappointing. You talked about large institutions and small institutions. Ironically, the people implementing drastic segmentation the most are the smaller institutions who can do an end-to-end job of working that through all the programming and delivery and all of their communications and education.

Large institutions are so focused on income generation that they are only one exhibition away from failure. They are increasingly employing commercial marketers who don't understand what the audience is capable of, or how people appreciate art and what it means to them. They are almost dragging the sector back to demographics.

We've seen a number of high-profile institutions snapping back to demographic profiling and targeting, and throwing away all the insight into why and how people engage, and just asking how can they sell tickets to local working professionals who are, say, between the ages of 25 and 39.

That is a trend.

KATE: It is about the why. Why are they conducting this particular study? Why is it that they want these answers? Sometimes those that have the most money have the most data, but all institutions are facing this abundance of data. The part that's often missing from institutions of any size, and of any flavor, is that they don't know what to do to turn that data into insights. The translation of the data into action can be the part where people get stymied. Much of my work these days is helping folks to build their capacity for data literacy, to help them see the data for what it can tell you and turn that into action.

KEIR: Kate, can you talk a bit about the similarities and differences you've observed working across the library and history sectors as compared to, say, art museums?

KATE: The big differences I see are the questions that each organisation type is asking. For example, libraries don't face the same community-based stigma that museums often do, certainly in the United States. U.S. museums have been notably elitist, and libraries are seen as open to all, providing programs that the community has input into and suits their needs and their wants. They are very different in the questions they are asking.

In zoos, a conservation message is key. A lot of my clients in zoo or nature center settings are thinking about motivation and what might they do to create behavior change or a systemic change around conservation.

Sometimes organisations collect data for the sake of it or to say that they have or to put it in a report. Sometimes they do it to make real change.

ANDREW: I think you're right, Kate. It's not necessarily about size. It is about orientation. The problem is most prevalent in those organisations that are financially focused rather than audience-focused. They tend to be larger because they have more to lose, but it isn't true of all large organisations.

My practical experience in the United States, Europe, Australia and New Zealand is that we're finding a correlation between those institutions that are closer to the community. One of the things that they have in common is that they're free to enter.

In London, where all the national museums are free-entry, in Stockholm and Australia, where they're largely free-entry, those organisations have been first in asking questions about why they exist and what role they have in terms of community and society. They've been quicker to move to being audience-focused and wanting to understand the value that they have in society and the difference that they make to visitors.

Organisations that are more dependent on earned income have a tendency to be anxious that whatever they do is more directly connected to sales. Those organisations are in real turmoil. Although a research institution or a conservation organisation has a social value proposition, they often have commercial management who don't have faith in the audiences or their ability to connect to that scientific, conservation, or altruistic message about nature and wildlife. They cling to the model of running an attraction where selling ice cream and cookies is given higher preference than engaging people with the core of the organisation.

A number of organisations adopt a motivation-centred approach to their audiences, where they invite the audience in to the core of what they offer.

KEIR: What sort of questions are being asked about digital in museums at the moment? Are they the right questions? What should we be asking that we're not?

KATE: A question that I hear often is the question of relevance. What is the value and role of museums in society? What does it look like at the neighbourhood, community, city, county, state, national and international levels? Some of the questions are: What do people take away? What is it that resonates? What are they thinking about a couple of weeks later? What are they telling their friends and family about? A lot of the questions are about revenue generation and getting folks in the door with a marketing focus more than an audience, visitor or community focus.

Probably my favourite questions are about the journeys that people have in our institutions and the questions that acknowledge that the experiences folks have in museums are part of, and not separate from, the rest of their lives. Perspectives and questions that are looking at intersectionalities and integration with the whole person, which is why segmentation or psychographic approaches and the use of things like personas, where museums are trying to think about the whole person, are encouraging.

If done in a way that's integrated that thinks about a person as a whole person – as a social being that interacts with other folks, their friends, their families, their partners, their communities – that's when the questions get really interesting. What I love about our work is asking questions and then figuring out how to answer them. If you're doing it right, you end up with more questions, and it becomes this wonderful, iterative cycle that can drive the work to beautiful places.

To me, it's the most generative and rewarding part of this work.

KEIR: That reminds me of a research workshop I attended recently where Erica Gangsei, the convener, said, "we're not here to answer questions, but to get past the obvious questions to more interesting, deeper, and harder to answer questions."

ANDREW: The kinds of questions that institutions want to ask and know the answers to are definitely evolving. But it depends who's asking the questions. The marketing department, for example, wants to know "Who are these people, where are they from?"

Increasingly I see the research moving up the food chain. The commissions or questions are coming from a higher level in the organisation. That leadership or trustees want to ask those questions reflects the increasing status of audience research.

It might be true that organisations only spend money on measuring what matters to them. They used to ask about delivery. Now they're asking about engagement. But I think the questions are now actually about relationships. What do we deliver? How are people engaging? How are they making meaning? What meaning are they making? How can we see value in the society? How can we engage people over a lifetime? How can we build that value? To be more audience-focused can be at the expense of other kinds of focus, collections or income generation or others.

I think that reflects the evolution of the sector.

KEIR: That sounds like the sector that I want to be a part of.

I want to talk about someone whose work I've utilised, the researcher and author John Falk, whose visitor-experience research shows how deeply personal museum visits are and how strongly the visit is tied to a sense of identity. How does Falk's visitor-motivation taxonomy differ from MHM's culture segments? What do they share in terms of utility?

ANDREW: There are many different ways of approaching trying to fathom audiences and understand them. Human beings are complex, so any system is not going to mirror the real world. There's a lovely quote . . . "All models are wrong, some of them are useful."

If an organisation starts to think about the differences between audience groups – why some people believe or think or behave in one way and others behave in another – the organisation is going to get a better outcome than an organisation treating everyone as a homogenous group, even if the tool they're using is crude.

A lot of these approaches are designed for a specific purpose. For example, there are systems focussed on how people learn, which are brilliant if you're looking at how people learn. There are systems that are only to do with how to reach new markets, which are useless when looking at how people learn.

There are systems looking at how people engage with interpretation or make meaning in gallery contexts. They're pretty useless at prospecting new audiences. John's identity system is remarkably similar to a system that we authored in the mid-90s, working with *Tate* in the U.K. which we called visit modes.

What that and John's system have in common, and what differentiates them from segmentation proper, is that they're not actually persistent segments. Instead, they are modalities. They are behaviour modes. Not only might you change the visit mode or identity on a different visit or subsequent visit, you could change during the visit. That depends on the circumstance, who you're with, how long you've got, where you've been and your level of knowledge. So, these are most useful for looking at visitor needs as they engage on the journey of a single visit.

Segmentation is based on fundamental belief systems and fundamental values. That's not to say that segments are fixed, because people may move, but the pace of change is slow. People arrive for a visit in that segment, leaving the same segment. When you go and have your next contact with them, they're in the same segment. These segments are persistent and are about how people see the world.

We have culture segments, which work across learning, product development, digital engagement, marketing and curatorial, and they work for visitors and non-visitors. It works with people who are not residents in the country so it's not reliant on European coding or anything like that. It works with tourists. It works with different cultures, so we've tested it in Islamic countries, we've tested in South America, we've tested it in Northern or Western countries. It's based on something that's deeply human.

These approaches all have value, but for different applications.

KEIR: I know I jump between motivational modes if I'm with someone, or on my own, in a museum where I'm familiar with the material or one where I'm not. Do you have a recent example of an organisation doing audience research, taking new information in and then changing direction?

KATE: I worked with the *ECHO Lake Aquarium and Science Center*, in Burlington, Vermont. They're a small science centre and aquarium right on the lake across from Canada.

Burlington is a university oasis that houses the University of Vermont. For years it was, compared to many cities, an overeducated area that was predominantly white. In the late 80s, it was designated by the U.S. government as a city that would welcome immigrants – a federal refugee resettlement program. As that happened, race and ethnicity demographics started to change quite drastically. Over the course of 10, 15, 20 years that became multiple generations of demographic change, yet this aquarium and science center wasn't seeing those changes in their audience demographics. They had questions that stemmed from that. Who are we serving, who are we not serving? Who feels welcome, who doesn't feel welcome?

They had acquired funding and were considering expansion to an adjacent plot of land. They knew this might be an opportunity to serve the needs of community members that they weren't engaging at that time. The New Americans, the immigrants, but also other folks. The homeless population had grown in Burlington and they were wondering what they might do to serve them. There were a lot of needs around early childhood education and around dealing with young people, both teens and college-aged students. Those were some of the targeted audience groups they were interested in better serving.

They brought me in to work with community-based non-profits that were already in close contact with those audience groups and communities, like a non-profit that worked specifically with homeless populations and another that worked with early childhood education providers. We brought these community service providers to *ECHO Lake Aquarium and Science Center* and had them do video walkthroughs. We started on the outside of the building and, using a documentary style filmmaking approach but with a modest budget, I facilitated a "think aloud" methodology where I held the camera on them, but below my chest so it wasn't in their face, and I asked them to say out loud about anything they were thinking in their head. I asked them to talk about what they heard and what they saw, who they heard and who they saw, when they looked at the outside of the building. What message did the building send? Who was coming down to this part of the lake front? We asked them to use all of their senses to take it all in, then to comment on what they saw and didn't see.

Then we went closer to the building and they talked about the messages physically on the building, the signage, the accessibility or inaccessibility of the spaces, the bike racks, the colours used, the landscaping, everything. Then we went inside and did the same thing with the exhibits. We ended at the new outdoor space because then it was a chance to take in everything they had seen or not seen and dream big about what this parcel of land might be able to be, or what role it could serve for the community they worked closely with.

We then did this directly with members in those communities that we were referred to from the service providers. We brought in folks from the homeless community, we brought in New Americans and we did the same activity. In the end, there were immediate changes that could be made based on the data we collected.

We learned through these video walkthroughs that it was very hard to know what was inside the building from the outside. The *Science Center and Aquarium* wanted their lobby space to be welcoming to anyone that was down at the lake front. You didn't have to pay to use that space or the wonderful gift shop inside, but you wouldn't necessarily know that. People didn't get that message. The first thing you saw as you entered was the marquee with all of the prices and stanchions that kept people from going farther.

They made an immediate decision to switch those around. To move the gift shop closer, right off the door. They made it so you could see exactly what was inside from the outside and brought some exhibit components into the lobby so that everybody could interact with some of the core content, regardless of if they had paid entry or not.

This quick decision sparked all of these other decisions that they began to work with the architects on. I think that is a great example of people asking the right questions, answering them creatively and then acting on them.

KEIR: That's fantastic. Andrew, can you provide us a corollary?

ANDREW: Yeah. If you look at the idea of prestige, which is part of philanthropy, many organisations are built to attract status. Many work on a hierarchical model of the more you give, the more prestigious that is, you move up a level, you get more rewards, you get more status, you get invited to more swanky events and activities. That assumes that social status and prestige was the driver that was fuelling gifting and donations.

We did a piece of profiling work for the *92nd Street YMCA*, which revealed a huge number of the "expression segment." What's interesting about the expression segment is that they're so uncomfortable with elitism that they would go out of their way to avoid exclusivity and seek inclusivity because the fundamental driver for them to be involved is to be part of something bigger than themselves. To support institutions that they feel are doing good in society, creating social value, growing social capital.

So, *92nd Street Y* ran a really intelligent campaign inviting people to explain what it was that kept them coming, what it was that they belonged to. They created a video, stories and still photography of people talking about their passion and holding signs saying why they were involved. They recycled that into a giving campaign that had a huge response and spiked their donations, particularly from the expression segment. It was an invitation to make common cause with the organisation rather than an invitation to climb some kind of social recognition ladder.

The second example I'd like to give is from the *Tate* in London. For a long time, they've wanted to broaden the types of people coming. They succeeded in making *Tate*, I think, the second-biggest visitor attraction in the U.K., which is astonishing for a modern art gallery. They've reached 5.8 million visitors, but those 5.8 million visitors still are not fully representative of the community that could be serving.

The problem, fundamentally, is that while identifying, say the African Caribbean in London, as a target group is perfectly legitimate to use that as a demographic mark-up to try and monitor year on year, it's really not a good way to target and engage that community. You can't say to the community,

"African Caribbean people, please come to the gallery." It doesn't work. What we've determined is the people most likely to come to *Tate* within that community, with substantive research, are again the expression sector.

By targeting expression segments within that community, they can make a bridge to the community because they're the group most likely to respond. It's easier to target them to look at the kind of programming activity, the kind of language to use, the kind of imagery to use, because culture segments help us to shape all of that.

Once you invite them to build the bridge to the organisation, expressions bring other segments across. It's understanding psychographically how people in what appears to be a demographically defined community can make contact with your organisation.

KEIR: I think back to when we were developing *SFMOMA*'s website and its on-site app. Understanding of the motivations and the relative sizes of the different arts-interested psychographic groups within the Bay Area was fundamentally important to our digital and content strategies. Kate, how can we take a more narrative-based approach when building digital experiences or tell digital stories?

KATE: I'm getting a lot more requests to do front-end and formative work on digital projects, either web-based or digital installations on the museum floor, which is wonderful. The same sort of systems that we talked about for non-digital or non-technology-based projects are applicable in technology projects as well. The ones we're using now would be those segmentation or psychographics or things that we might borrow from other fields like anthropology and ethnography and design, by doing design thinking or usability testing. When you pair those with what we already know within audience evaluation, those methods can be even stronger.

There are many complementary approaches. I'm seeing "museum evaluators" or "audience researchers" going outside of our own disciplines to get more creative about how we approach those front-end and formative evaluation projects so that we can really get to the heart of what various audiences, various people, want out of our products and our services.

If done right and early enough, front-end and formative evaluation can help to avoid missteps along the way, and it cuts down on needless and often costly expense. I've been really encouraged to see how many of my digital clients want to engage in those earlier steps. There's so much promise there.

KEIR: Around the reopening of *SFMOMA* in May in 2016 we had a pretty ambitious and robust digital strategy with a lot of interactive elements and interactive spaces.

After we set our expectations, we realised that we didn't quite have enough money to include formative evaluations, summative evaluations and that first year of changes that you need to make when you discover how things are being used. We took a tough decision to kill an entire project and funnel the funds from that project to the others.

This meant that we could invest in the rest of the projects with formative and summative evaluations and leave enough money to make six months' worth of tweaks based on use. In the end, one of the best decisions we made about digital at *SFMOMA* in the lead up to that grand reopening in 2016 was to do fewer things, do them better, and with an acknowledgement of how they'd be experienced and then respond when you get those things wrong. No one gets it all right. That's not a thing.

KATE: That's a really good point. But there can also be a false sense of stages in museums. Summative in museums is hopefully never truly summative.

With your project, I was really impressed that you had built in that time and those resources to acknowledge that the summative wasn't going to be a bookend. It was going to bring up a new series of questions and new opportunities. The way that I always think of it is well summed up in a quote that I use all the time from an evaluator I really respect named Michael Quinn Patton: "Research is to prove, and evaluation is to improve." Thinking of evaluation as iterative and as a cycle is so helpful, rather than thinking that it means you're done.

KEIR: Andrew, how should museums be thinking about including ideas of engagement or affinity, even belonging, when researching user experience of their digital products? Is it possible for someone to have a sense of belonging from a website?

ANDREW: I believe so. Broadly speaking, the outcome of UA (user acceptance) is "this is usable" and broadly speaking the outcome of UX (user experience) is "this is useful." Whilst there's a lot of good in testing UX, most user-experience testing is about efficiency. How quickly, seamlessly, does this website, app or interactive in the gallery deliver this experience or this knowledge? It doesn't necessarily take into account how effective it is at things like engaging the visitor. Or, meeting the visitor's deep-seated needs. Often, it's about a push: What can we give the visitor? How quickly can we deliver it to them?

So, we go beyond user engagement to look at how users feel about that engagement emotionally. How is this interaction or content meeting the user's needs? The outcome still might be that it's useful, but it might be, "I love this exhibit. I love this app. I love this experience."

We go still further to say, "How does that speak to their relationship to the institutions? To what extent does it make them feel part of the community? To what extent does it make them feel differently about the organisation? How relevant do they think the organisation is to them, their lives, their community? Do they feel the organisation is open, transparent, porous, caring?" All of those things that form relationships between individuals and communities and institutions.

The outcome of that is not just this is useful or "I love it." It's, "I love the organisation, or I feel part of the organisation or I feel aligned to the organisation." That's user affinity. We do this string of four things: user acceptance (usable), user experience (useful), user engagement (I love it, it meets my needs), and user affinity (I feel aligned with my organisation. I belong). You can get that from any kind of experience including digital.

KEIR: Right. In that understanding, we could end up recognising some audiences over the needs of others, especially who have fewer digital interactions or fewer digital skills or fewer ways of being measured.

Is that something that you hear your clients worried about? That through measuring certain sorts of people you end up with a skewed vision and potentially skewed actions?

KATE: There's always this risk. By focusing on one audience, whether that be for programs, building expansions, exhibitions or museums, you always have the chance of excluding, ostracizing or marginalizing other audiences. I don't think that's any more or less true for online and digital audiences than it is for any community or audience.

Museums always make choices with their data. You prioritise who you listen to and who you choose to hear and who you connect with and truly want. You choose which data you analyse, which of it you care about, think about and ultimately act on. I think it's a fair concern to wonder if we are elevating the visibility of online and digital audiences, if that means that we are going to override less connected audiences.

It's definitely a risk and it's one that, as a field, we should continue to talk and think about how we're going to address it.

KEIR: I also think about where the data is coming from. Most institutions have pretty good idea about their member base because they've had interactions with them, more so than for non-members or non-supporters. When you're taking the segmentation approach and end up with a lot data about people who can't afford a membership or to financially support a fee-paying institution, do you end up with a vision of your audience that may not match what your actual audience is?

ANDREW: The danger is holding up a mirror and only seeing ourselves. Our ideas become self-fulfilling. There is a danger that we become data-focused or data-centred, or data-driven, only using data that's there and available.

I've seen large national institutions amassing big datasets from mobile use, and assuming that they're getting a picture of their visitors. But we know from very carefully sampled studies that the digital picture that they're telling is nothing like representative of a real audience. Even if they're collecting it from their existing attendance, it's potentially very skewed. One of the things we worked on with culture segments is the need for a system that works equally on non-attendance.

I do worry big data is quite norming and quite fulfilling. We think about data as being cold and factual and zeros and ones on a spreadsheet. But we're equally talking about human responses. And the human evaluation and even the humanisation of research is our best friend. We don't ever just look at spreadsheets or automatically collected data from Google Analytics and think that we understand the audience.

There is always a place for looking right into the eyes of visitors and non-visitors and making a human connection, because that is the only way to understand the rest of the data. Most of my work swings violently between fancy, cutting-edge, creative technology-driven methodologies and talking to people and watching them and listening to them. It's only by doing that that you make sense of anything.

KEIR: There's two really important threads in what you just said, Andrew. One is that the digitisation of the data is potentially dangerous or misleading. That technology like Google Analytics or heat maps of building usage shouldn't act as a stand-in for understanding your audience by doing the in-gallery observations and talking to people.

But you're also implying that you don't need any technology. The best technology is asking people, just getting an audience in a room and asking the questions that you are, as an institution, posing. Is that right?

ANDREW: Yes.

KEIR: What about small to medium museums, who may not be able to afford surveying? What could they do today?

ANDREW: I'm addicted to one-to-one or small groups of people and listening to their reactions and stories and getting in dialogue with them. The data generated by measuring and monitoring things are useful to help identify gaps or focuses on problems, however the real meat of my work is done in hand-to-hand fighting. It's close, careful listening. I want to humanise research before we get carried away with our own methodologies.

KATE: When I was still working internally at a museum, we were the only museum locally in Denver that had evaluation staff and had a department, albeit small, focused on evaluation. We quickly learned that all the institutions in our greater metro area were hungry to learn about evaluation and to build capacity. That was true for the two staff and one volunteer at the tiny historical site in the mountains, as well as for the giant institutions in Denver.

Working together, we learned that often it is those face-to-face conversations with visitors – where you don't a need high-level evaluation skill set – that are so informative. That's not to say that rigorous, thorough evaluations, or bringing in evaluation professionals, is a waste of time or money. But it begins with the simple. It begins with what folks hesitate to do because they feel they might not have the experience. We all have experience being human and having conversations. It really can start with that simple and very straightforward approach.

KEIR: Andrew, you talked about being vision-led, but also relentlessly audience-focused. How do you balance those mission and curatorial and artistic visions with audience interests, identity and expectations? Are museums less ambitious or less adventurous, even less academically rigorous, if they're driven by audience motivations and not that mission, not that artistic vision?

KATE: I've been really encouraged to see a lot of my clients go towards asking questions about non-visitors and potential audiences. Not just in a way that I used to hear in terms of what's the potential of getting people in the door, but understanding their place in their communities and their function as a part of those communities.

That puts humans at the centre and understands that we're part of an eco-systemic world. More of my clients want to understand how they can be part of, rather than separate than.

I found that my clients that incorporate audience and visitor voice into their mission and vision end up with a far more compelling mission and vision. Perhaps curatorial vision or strategy as well. For example, I'm working with a museum on their brand strategy and marketing. Their brand guidelines and what they hoped that visitors would say as they left the building used flowery and bold language. We did a study asking visitors as they left to describe what the true values of that institution were. Then we had visitors give us the reflection of those brand standards and values and they were far more reflective of what the institution truly is. You can do that for mission and vision, and your curatorial agenda.

The audience should be considered one of those groups that we need to work hand in hand with if we want to do our best work.

ANDREW: We were doing audience forums for one of the world's biggest art museums, bringing the audience together to talk about their experience. We had the staff and the director, and everyone present in the room. It was a fantastic session exploring people's fears, hopes, ideas about art, about the organisation. At the end, the director said, "that was really useful and really interesting, but where did you find them?"

We said, "well, they're your audience." And he's saying, "but they were so intelligent, so articulate and interesting, and they ask such good questions. How did you choose them?" We said, "they're just your audience."

This surprise that the audience are intelligent and capable of engaging, that they may not have a lot of domain knowledge or experience, but they are curious and capable of understanding things, tells you what one of the big issues is: the assumption, particularly in art museums, that only a certain portion of the population are capable of appreciating art, and if you want a bigger audience then you have to compromise artistic integrity in order to make it understandable or palatable. How much of your artistic integrity will you give up in order to get a broader or more diverse audience?

We've spent a lot of our professional life trying to disabuse people of that idea. All that's required is to understand people and their communities and where art fits or doesn't fit into their life. How it might be relevant? How might we make it more relevant and engage them on their own terms and explain our knowledge in a way that makes sense and provokes their curiosity? You can have a mass market, deeply engaged with art and have the highest integrity. That, to many people in the sector, appears to be squaring the circle. They can't bring themselves to believe that that's possible.

That's why they see marketing and audience research initiatives as almost a threat to curatorial integrity. That's one of the issues we're grappling with.

We got tired of how generic mission and vision statements are. We took 10 mission and vision statements, put them around a room at a conference, gave all the delegates the list of all 10 organisations, and ran a competition to see if people could match the organisation to the mission statement. The winner got two out of 10.

Mission and vision statements are borrowed from business. Business plans are written by people are trying to run the world more efficiently. We're trying to encourage people to think of a cause statement and write a manifesto calling people to make common goals with them. You can't change the world unless you bring people with you. The audience and the community are inexorably at the heart of what you're doing. This is where we came up with the idea of unequivocally artistically or vision-led and relentlessly audience-focused.

KEIR: If you have a microphone Andrew, you should drop it.

KATE: If institutions could focus more on their core values and their calls to action it would be far better time spent than on mission and vision. I've done that same activity in training and workshops, where I will present people with mission and vision statements that aren't from museums, that might be from other non-profits or even Fortune 500 companies, and they still think it could be one of their museums. We all use the same jargon and language and buzz words.

KEIR: Understanding a museum's community, its current and its potential audiences – their desires, their motivations, their self-identity, their narratives and their language –should be a prerequisite for museum strategy. Could you summarise how you think museums could prioritise which of these audience-focused strategies to deploy, or how to mix them together so that they can feel that they've got a landscape, or they've got the understandings that allow them to go forward as an audience-centred organisation?

KATE: The biggest step that I've found for most of my clients is understanding that there are a lot of myths that need to be debunked about evaluation. There's this assumption that you have to be a mathematician or a data geek, and we're going to analyse data in a way that dehumanises the process and dehumanises the audience. We want to do the exact opposite. When we keep the individual and communities at the centre and elevate those voices and prioritise them, that is when we're doing our best work.

It can start with getting on the floor and talking to visitors and understanding their perspectives better. It can be reaching out and talking to them where they are, both inside and outside of your museum, then recognizing that there are incredible resources, many of them free, for building our skills in these areas.

Everybody is an evaluator. Everybody asks questions. We all feel best when we're connected to one another. Most of us in museum settings love learning and love connecting. We might be artists or curators or scientists, but we're working in public settings and there's a reason for that.

It's then figuring out ways to scaffold to get you to the next step. Sometimes it's bringing in outsiders when you need that extra boost. I'm always working to build evaluation capacity with my clients, so that after I work with them a handful of times, they only need to bring me in for certain studies or certain boosters that might be out of their immediate expertise. My hope is that when we end a project, you've learned some new things that you are ready to keep doing. That evaluative culture becomes part of your institutional culture and becomes as important as all your other core values and your mission at your institution.

ANDREW: Well said, Kate. I agree with all of that. I have yet to write the PowerPoint presentation that is more powerful than senior museum leaders hearing firsthand from the audience. There is no substitute for that human impact.

I fear that we may have created a priesthood of evaluators and researchers who are the only ones allowed to speak to the visitors and reveal the insight to the museum professionals needing the information. If all we do as evaluators is break down the mystique of evaluation and make people feel that they could talk to people or watch them, we're going to end up with a more visitor-focused museum sector. We can give training, we can build capacity, or we can come in and help tackle those knotty challenges, but this will only work if we embed this as everyday practice in museums.

My killer question is who in your organisation is responsible for curating the knowledge that you accrue from evaluation? What's the mechanism in your organisation for looking after that precious knowledge? For curating it and applying it to projects?

KEIR: Good questions to close out with. Thank you both so much. This has been a great discussion.

This conversation acts as a timely reminder that gathering data is meaningless if you, or your organisation, are not going to use the data to inform decision-making. What is measured is a useful proxy for what is important to an organisation, be that clicks, visitors, daily revenue, social followers, school group attendance, cited publications, or the happiness of visitors and staff. As organisations try to differentiate between understanding motivational modes, socio-economic backgrounds, psychographic segments, visitors vs non-visitors or other strategies, in most cases, as both Andrew and Kate note, the best way to start is to simply talk with your audiences and actually *listen*.

This discussion goes beyond the necessary growth in data literacy within the museum sector to a future where formative, summative, qualitative and quantitative evaluation is part of everyday practice. As very few organisations have full- or part-time evaluators on staff, it's incumbent upon creators of public-facing content, experiences and programs to also be evaluators. As Kate Livingston notes, "everyone is an evaluator, everyone asks questions." The key challenge is what you do with the answers.

For more on evaluation and measurement, we suggest you read:

> **Conversation 2: LaToya Devezin + Barbara Makuati-Afitu**
> **Conversation 4: Sarah Brin + Adriel Luis**
> **Conversation 11: Tony Butler + Lori Fogarty**

For more on the changing definitions or understandings of culture, we suggest you read:

> **Conversation 3: Lara Day + David Smith**
> **Conversation 4: Sarah Brin + Adriel Luis**
> **Conversation 5: Sarah Kenderdine + Merete Sanderhoff**
> **Conversation 10: Arthur Cohen + Tonya Nelson**

CONVERSATION 9

Brad Dunn + Daryl Karp

"Digital for me is a language. It is not just a technology."

– *Daryl Karp*

FIGURE 9.1

Brad Dunn (U.S.) brings a mix of digital and marketing agency work, his journalism and theatre production experience, and five years as Director of Marketing & Communications at the *Metropolis Performing Arts Centre* to his work at the *Field Museum*. At the 2017 Museum Computer Network conference he presented "Museum Digital Content as Journalism? An approach to original content," explaining the *Field*'s approach to presenting scientific information in post-fact America.

Directing *MoAD* is Daryl Karp's (Australia) first museum role in which she builds on more than 20 years in senior management positions in broadcast and digital media, most notably as CEO of *Film Australia*, and a Head of Science and Documentaries and then Head of Television Factual Programs for the *Australian Broadcasting Corporation* (ABC). Long fascinated by technology as Head of the Science Unit in the early 90s, she oversaw the first ABC website.

We brought Brad and Daryl together to discuss how cultural organisations can present as media companies, employing broadcasting or journalistic principles and strategies, and how these practices can affect how people experience a museum visit, how they engage in-person and online, and the trust they place in museum content and institutions themselves.

This conversation was recorded on March 9, 2018. At the time, Brad Dunn was Web and Digital Engagement Director of *Field Museum* in Chicago, U.S. and Daryl Karp was Director of the *Museum of Australian Democracy* (MoAD) in Canberra, Australia.

KEIR: The idea of a museum as media company – by which I mean a content-producing, multi-platform organisation that happens to have a building, exhibitions and events – will begin our discussion. What commonalities and differences between museums and media companies or broadcasters do you see as pressing right now?

BRAD: At the *Field Museum*, I've been looking to traditional media – journalism, newspapers, TV – to see how they have built their relationship with commerce. This is pressing, as certain areas of finance dry up and others become more available.

I joined the *Field Museum* so I could serve an institution that produced content that was of societal value, that was mission-based, and had an integrity to it – where it wasn't up for interpretation or spin. It was scientific research. That's run head on into this idea of branded content and advertorial content as ways of supporting the people who fund and support the museum.

I've built a wall around our content operation, and I'm working hard to keep it that way. There's got to be a strict divide. So, one of the issues that we're struggling with is the relationship between content and commerce.

DARYL: I come from a public broadcasting background. I think commonalities between the values of a public broadcaster and the values of the museums to educate, entertain and inform are hand in glove. I've taken almost all of the processes that I used as head of factual programs at the [Australian national broadcaster] ABC to change how we approach our storytelling and our exhibition and event experiences at *MoAD*.

We're a very small organisation and being nimble and responsive is critical. We have implemented what I'm calling a TV model in the museum. I have the commissioning editor of content, who sits across digital, schools, learning, exhibitions and content development. We have executive producers for each exhibition. We have a matrix-style operation, rather than silo-based divisions. I often fall back on a three-act narrative structure for exhibitions or digital experiences, and – drawing on the independent producer model – we're continuing to refine a commissioning model that captures great ideas from creative thinkers who may not be schooled in museology but know how to tell fabulous and engaging stories.

The other thing that I've introduced is a budgeting and editorial process that makes it easier to commission and frame thinking – the museum-equivalent news story, a weekly magazine item, a half-hour documentary and a blue-chip type experience. Each of those have a different production framework, a different budget associated with it and a different time frame. We can put something together in a couple of days. Ideally, our response rate, even for our bigger exhibitions, is six months. Even less, if I have my way. That's been another big breakthrough. Because [being] timely and relevant is a big thing for me.

The digital team is engaged from day one, incorporated into and influencing the broader exhibition experience. It is before you get into the exhibition, in the exhibition and after you leave. For an exhibition that's currently in development – Democracy. Are You In? – I've given the exhibition team the brief to come up with an entry experience that is, effectively, a physical manifestation of the digital experience. We know what our audience wants in the digital space. They tell us. We can see what they respond to. So, what does that look like if we take the digital experience and convert it into a physical space? How do we give it the voice, the energy, the dynamism?

For me, the digital experience is both on-site and off-site. I'm trying to make it a seamless experience. Brad, does that make sense to you?

BRAD: Very much so. Digital for us is not just channels of broadcast. There's a community that we manage and engage. The community continues to exist when they're in the building. That doesn't mean I want to create "eyes on phones" experiences. But I do think about the ramifications of the fact that everyone does have this thing in their pocket when they're here. How can it add to the experience of visiting the museum?

DARYL: We've had great difficulty delivering transformational experiences online. We're finding that in the space of democracy, people feel more comfortable being anonymous. We want to have civil, engaged dialogue, but to find that space where you can have those sorts of conversations often requires people to have a relationship first. That's our big current challenge, because politics is polarising. How do you have a community conversation between people who don't necessarily know each other?

For example, every on-site exhibition has a call to action. When we use digital calls to actions, the response rate for digital versus analogue was about 15%. When we ask people to tell us what they think, handwritten and anonymous, people would write beautifully, well-considered, relevant and timely commentary. But they would not do that online. Even with the personally identifiable data stripped out.

Once a year, we do a big, collaborative art build. In 2018, it's the *Card Castle*. We've taken seven key values that underpin a strong community – hope, courage, community, diversity, trust, kindness and curiosity. We invite the public to choose a card, and write a story of hope or courage, which we then build into a shared castle experience. We're open 'till 11 o'clock for four nights over two weekends during a big festival in Canberra called Enlighten. This is the fourth year we've run it. I expect we will have 25,000 participants over four nights. We haven't yet worked out how to have that same transformative and meaningful communal experience in the digital world.

BRAD: It's interesting you mention the communal piece, Daryl. We have research here that shows that for many people the social experience is one of the most important elements of the museum visit.

My career began in theatre and, for a while, I was directing. While I was, I started improvising. With improvisation, the audience loves that they are seeing what's happening only milliseconds after the performer themselves figured it out. So, part of the joy is the shared experience of discovery between the performers and audience. Coming with that context into the museum world, I started to notice that we spend a lot of our adult lives being told that we're supposed to have certain pieces of knowledge already, or that we're supposed to have answers. But then people come to a museum, and it's a safe space to experience the feeling of being curious, to have questions about things. Doing that with other people creates a communal experience.

As much as I'm a digital devotee and a technologist, I always want us to ask hard questions before we bring something digital into the actual space of the museum. I want to enhance that experience, I don't want to disrupt it, because that communal piece is really important. That's a fundamental human thing, of discovery, learning, curiosity and being able to do that with other people.

DARYL: Brad, we're on the same page on that one.

KEIR: What is an effective way of attributing valuing that digital engagement or that physical-digital engagement? If transformation is really important, how is it being valued and measured?

BRAD: We are deep in the throes of trying to figure that out. For me, the key performance indicator here is, "Are people feeling curious?" Are they feeling engaged enough to have a personal experience that they take with them when they set down their phone, walk away from the computer, or leave the building? That is a thing that I desperately want to figure out how to measure, because I think it is built into our mission.

The actual craft of science is creating hypothesis. And hypothesis comes through research, but it also comes through dialogue and being open to test different ideas. While there are certain incontrovertible facts, as we're trying desperately to remind the American public, there's also room for people to engage us in conversation. These are not broadcast mediums. It is a community and they have voice in it.

DARYL: I agree. What is your vision or your thinking on digital? When I think of broadcasting, it's simply a mechanism to communicate. I'm trying to introduce this idea of seamlessness – that it is a single, unified experience that you can access however you wish.

BRAD: I look at it as a circle, where we are conducting research, we're producing content in various forms from the research paper to the social media post, and then the audience gets to give us feedback. I like to think of it as giving people opportunity to jump in anywhere on that circle that they're comfortable.

Some people don't want to talk with us. Some are happy to tell us everything that they think. And you know, there's a couple of people that, anytime we post about our work in the Amazon, or any of our work on conservation and climate change, they'll always tell us we're wrong. And we never, ever, silence them. I did have someone come to me once saying, "Why don't we just hide those comments? Why don't we turn that off?" And I said, "Because that's not democratic." We're not going to say they're wrong, we're just going to say, "Here are the scientific facts. If you want to engage us with science, we'll hear that." But we're not going to try to silence people.

Some people don't want to engage in any of that. They just want to consume. They want to read what our point of view is on things that are happening here in Chicago, which is part of our museum's work here in local communities. Some want to have a more full-fledged kind of experience. The thing we're trying to figure out is how to bring that into the galleries.

It's easy when you have an exhibition like *Tattoo*[1], because so many people have tattoos. All you've got to do is ask someone to talk about themselves and they'll happily do it. We generated millions and millions of views and tons of interesting conversations about people's personal lives through the story of the ink they've put on their body. That was relatively easy. How do

---

1 The *Tattoo* exhibition was developed and produced by the *Musée du quai Branly – Jacques Chirac* and was presented at the *Field Museum* in 2017.

we actually get people to think more deeply? I don't think we've cracked that nut at all.

DARYL: Where does digital sit in your museum, curatorially? Is digital an equal voice? Is it part of the conversation from day one?

BRAD: When it comes to things we want to publish online, we have final say. Sometimes curators come to us with the thing they're working on. Sometimes, we have to go to them and say, "Hey, Gary, haven't heard from you in a while. . ." Or, "I just saw that you got this grant approved, tell me about that work that you're doing." But when it comes to the exhibition, we're often considered, rightfully, in a secondary way. The content that they want to focus on really comes from the curators and exhibition development staff.

I like to say that through digital, people can see the museum they can't see when they're here. There's all this science that's generated in an exhibition that people will never see, so we try to put that out in social and on the web.

DARYL: What you've got is almost the opposite to us. You have these gobsmackingly wonderful objects. We have —outside of the exquisite building, which was the original Parliament House – the fact that history happened here. So, I'm much more able to explore big-picture ideas, where digital is often the vehicle.

Digital for me is a language. It is not just a technology. Digital is a different language and conversation that we have with visitors that adds value to the objects. It allows us to have a different kind of conversation, both in the museum and off-site.

BRAD: That's really interesting. I will do anything to try to steer people away from thinking about technology. It's really about design, content and engagement, regardless of the channel. People will come to say, "Oh, I've got to put this thing out on social." And I'll say, "Wait, wait, wait. What is the thing? Let's talk about it." Often it turns out it's not right for social media at all, but it's great for another purpose.

DARYL: Yeah, yeah.

KEIR: [At *SFMOMA*] we often fight to clarify "What is the experience you're trying to create? What is the thing you want people to leave with? What's the transformation you're hoping to achieve?" Our goal is to understand who the audience is and what their needs are, and what the project's goals are, what experience we're trying to manifest and what resources we've got. Then we see if digital is necessary. Because if you can do it without technology, if you can do it in a more human way, in most cases, you're better off.

I was reading something you wrote recently, Brad, where you said, "Don't silo your digital engagement efforts from your more traditional customer service people. Everyone in the organisation is in charge of engaging your audience." (Dunn 2017) Can you talk more about this idea, that digital is a practice and not just technology?

BRAD: Change in the sector is hard. It's not about a piece of technology or an area of expertise, but it's about a knowledge set that spans areas. The museums that will be successful will be the museums that figure out that it's not about these different disciplines but about how they work together and –

DARYL: And how they adapt.

BRAD: And how they adapt, right. To bring it down to Earth, we had [a visitor from the U.K.] who came to the museum and was very disappointed they couldn't [see what they'd come to visit]. Then they went to the store and they couldn't get what they wanted, and their kid was upset and they told us about it on Twitter. One of my social media people, Caitlin Kearney, said, "This person is so sad. Do we have budget to buy this dinosaur in the store and FedEx it over to the U.K.?" And I'm like, "Absolutely. Just do that." Someone in finance said, "Wait, why did we FedEx a thing from the store to the U.K.?" It's not really in our line of work. And I said, "This person was really disappointed . . . that's part of community management." And that's not really in the job description if you're doing social media or digital. . . . But it's about putting people first.

DARYL: That's a big one.

BRAD: It's hard for museums to do.

KEIR: How important is listening? Genuine, attentive listening for a museum?

BRAD: Oh my god, I think it's everything.

DARYL: It's absolutely critical. People feel very voiceless at the moment and very unheard. They want their contributions to count. Within our exhibition *Does Your Voice Count*, we asked people a series of questions. And the biggest response we had from the public when we followed up in the research was, "What are you going to do with our comments? We want the government to know how we feel."

I'm thinking of approaching the chair of The Standing Committee for Electoral Matters and seeing if I can do a presentation once a year on the issues that are being put forward by our visitors.[2] To be heard is really important.

KEIR: This could easily be read as a direct response to the declining trust in institutions, especially in the U.S. (Democracy Index 2019; Edelman Trust Barometer, 2017)

DARYL: It's not just in the U.S., it's across the board. Trust across the globe is declining.

KEIR: There is an increase in disagreement about facts. There's an increase in the relative volume and influence of opinion and personal experience over fact, and a decline in the trust of formerly respected sources for factual information. Museums are not neutral, and as institutions I believe they have a stake in this debate.[3]

---

2 The Parliamentary *Joint Standing Committee on Electoral Matters* conducts inquiries into matters referred to it by the Australian House of Representatives or a Minister of the Australian Federal Government.

3 The phase "Museums are not neutral," the social media hashtag #MuseumsAreNotNeutral and the T-shirts with the slogan all originated with LaTanya Autry and Mike Murawski.

Something that I found surprising in the 2018 Edelman Trust Barometer is that trust in journalism went back up, whereas trust in politicians, government and banks continued to erode. Brad, what can museums learn from journalism? How do we respond to this threat? Or this opportunity?

BRAD: I'm so glad that you say that we're not neutral. I know that that sentiment is very popular among museums' digital staff. But colleagues do talk about us being neutral and I say, "Come on, we're really not."

Yet there is research that shows that there is the chance that the more that we take a role as advocates, we risk sacrificing our role as experts. There's also research that shows that we have more leverage than we think (Dilenschneider 2017). We're re-launching our brand right now, and we're shifting towards showing people that we are a scientific institution. We're embracing the conservation work that we do, which a lot of people don't know about. So, we're about to dive headfirst into figuring out how far can we really go before some people view us as "a bunch of liberal elites, or people who clearly have an agenda."

I find that difficult because it's not that we have an agenda, it's just that we look at facts, look at research and we look at the truth. If the truth leads to one side of the conversation, so be it. We can't deny that. But we don't want to alienate people. We are going to be working through this over the next few years at the *Field Museum*.

DARYL: Leading an institution encouraging civic engagement and democracy, my message is not about who's right or who's wrong, in terms of the politics of left and right. We're non-partisan. Where I draw the line is anything that is racist or against the law. We will not put forward points of view that don't support a civil and civic engagement.

Australia is one of the world's leading democracies. It's on the decline in terms of trust and other measures, but the visitor, the audience, the voter, is part of the solution. There's no doubt that when politicians go through the museum, they look at it through the filter of their own biases. So long as we are factually accurate, the facts speak for themselves. One thing I do think is quite important is this idea that you raised, Keir, about fact versus opinion. That's probably one of my biggest concerns, the blurring of those lines. One of my personal passions is to get opinion highly labelled, to give it a different colour so everybody knows when you're seeing opinion. Whether it's in a museum, whether it's anywhere, you know that that is an opinion and not a collection of facts. That news is different to opinion.

KEIR: It's interesting to place museums in that context and disambiguate our various competing narratives that lead to all sorts of different interpretations of the world. That complexity and nuance is really fascinating, but when you have institutions that trade in reality, trade in factual information, shepherding your audience through in a way that is empowering and not disingenuous is really tricky. I think it's one of the great questions that we're going to have to answer as a sector.

DARYL: One of our big questions recently was around marriage equality, which was an ongoing and complex conversation about the human rights of a group of Australians who were unable to marry.[4] There was a lot of conversation internally about how to address it. There was desire from people inside the museum to be part of this momentous experience, [but] I felt that it wasn't our role as a museum to advocate one way or the other. We could comment on how the process took place, what the background and past political histories might be, where different groups had actually fought for change and how it had been brought about. But while the debate was going on, that was not our role. We would respond if anybody said something incorrect – about history, about democratic process or whatever, but primarily we facilitated a conversation to by others.

At the same time, we gathered all these outstanding objects, which will feed into an exhibition. The bill went through late last year. The first marriages took place early this year and our exhibition will open in June.

For me, the important thing is that we are one of the few really trusted organisations.

BRAD: Yeah.

DARYL: So, when we say museums aren't neutral, we have to bear in mind that we need to be trusted. Trust is our currency, but it doesn't mean not having a position.

BRAD: I agree with that. We really are not neutral. Yet, it's not our role to comment on social issues. We've created a document to govern ourselves that outlines the kind of things we can talk about that make sense to our mission, the kinds that don't. When it comes to things like what the EPA (Environmental Protection Agency) is doing, making terrible decisions here in the U.S. with regard to environmental regulation, it's not our place to say that. I do worry about the extent to which we appear to be doing combat with them. Do we risk losing our role as experts in favour of being able to advocate?

There's a certain momentum here now because of our brand relaunch, and our new, restated mission. But if we lose that trust, it doesn't matter. . . . Then nothing else will matter, essentially. So, I agree with Daryl. That is the currency we have to trade in.

DARYL: Absolutely.

KEIR: I think for almost all cultural organisations with a visiting public, there's continuity, connection and trust. It takes, in some cases, hundreds of years

---

4 In 2017, the conservative Australian federal government led by Malcom Turnbull created a plebiscite (a voluntary postal referendum) to give Australians a chance to vote in favour or against marriage equality for gay and lesbian Australians. The survey questions was, "Should the law be changed to allow same-sex couples to marry?" There was an overwhelming strong vote in favour of marriage equality, however many marriage equality supporters were critical of the process, viewing it as a costly and unnecessary delaying tactic by the then government.

for institutions to build that trust, but it is very easy to lose. You can lose that trust through one action.

BRAD: That's correct.

DARYL: What worries me about the notion that museums aren't neutral, is the corollary, that museums are biased. I think we are neutral. We fall on the side of facts or we fall on the side of research. To say museums aren't neutral, I think there is a public risk at a time where people don't know who to believe. And for museums to be actually walking around publicly saying, "Oh, well, we're not neutral." We know what we mean by that, but in the public sense, that's not what they will take away from it.

KEIR: Some of the trust museums have earned comes from years of making educated, research and scholarship-based choices about what to show and how, but they're still choices.

BRAD: I'm reminded of an experience I had a few years back. I was standing on the floor of our main hall and a couple approached me, and they said, "We don't have a lot of time today, what would you recommend that we see?" And I said, "Well, it depends on what your interest is. I'm a big fan of *Evolving Planet*, that's our exhibition that takes you through the history of the Earth and that's where we have dinosaurs. There's a lot of great stories and a lot of great learning in there." And they stopped me and she said, "Oh, no, no, sweetheart. We don't want to see any of that evolution stuff. We don't believe in any of that, but what else do you have?"

And I was like, "Well, sure. So, we've got a great hall on ancient China, and we've got this, we've got that, et cetera." And they said, "That sounds great, thank you so much." And they went on their way. I was utterly floored by this idea that they could come to a museum that they know espouses this idea that they're against, and just want to filter that out. They could still have a good time. There's still something about the museum experience that they took their time to come and see.

To know that we follow a scientific process, but to be able to filter out simply the stuff that's inconvenient for you, I thought, "That would be such a much easier way to live life, but I just can't do it."

DARYL: But I think that's one of the challenges that digital provides. You can filter out anything that's uncomfortable or that you don't like and only see what reinforces your position. I set out every quarter to clean up my social media feed. I follow journals and people that will give me a whole range of views. I do that to make sure that I'm getting different perspectives that don't reinforce my position. I even read what the evolutionists are saying, because I think you need to know what's being said in order to come up with the strategies that allow you to respond in a way that will nudge them slightly. I don't think you'll ever change people's minds when they're so hard core.

You know, belief is different.

BRAD: That's right.

DARYL: Belief just locks into an emotional position. That filtering is one of our big challenges that we have to deal with as a society, both digitally and even in the museum space.

BRAD: Yeah.

KEIR: Earlier, we spoke about content creation and distribution. Very few museums have created genuinely successful media content like the *Field Museum* has with *Brain Scoop*, especially when you compare them to broadcasters. What's holding museums back from making material that really scales?

DARYL: It's partnerships.

KEIR: Tell me more about that.

DARYL: Again, it depends on who your audience is and what you want to communicate with them. If you want to access a media-type audience, you need a media partner. Because they've got their own language. They know their audiences as well as we know our audiences. And the two audiences are slightly different. It's about marrying the two. And sometimes it's about having two different audiences.

The ABC was doing a series of interviews with all the living prime ministers, so I asked whether the journalist doing the interviews could include an additional question in exchange for permission to film in the *MOAD* building. The question was, "What was the positive impact of . . ." and they listed all of the prime ministers. Each of the living prime ministers commented on the positive impact of the other living prime ministers. We cut that together into a short clip that sat in the ante-room to the prime ministers' gallery.

We also partnered with the ABC for the full TV series. Millions of viewers saw our building and the prime ministers' wing. That series had its own language that was appropriate for television, but we were co-branded. We had access to that footage and were able to make sure that some of our needs were addressed in terms of minor narrative elements. And we got access to a body of work.

That's the sort of partnerships that we're doing. And we do that quite regularly. Do I think any of those [audience members] will actively come into the museum because of it? A small number, but that wasn't the purpose. The purpose was for us to have our purpose and mission and messaging reaching a different target audience group.

KEIR: Interesting. Brad?

BRAD: So, the *Brain Scoop* is our YouTube series that's produced by Emily Graslie. Her official title is Chief Curiosity Correspondent. She started the series before she was at the museum, with Hank Green. The museum wisely bought it, hired her and moved her to Chicago. She's a museum employee, who runs the show fairly independently within the museum and it's wildly successful.

Emily is an extraordinary personality on camera. She takes deep, super nerdy science and makes it fun. She genuinely loves it. And that enthusiasm comes off on camera. If you look at one of the other areas of big success for us, which is the Sue the *T.rex* Twitter feed,[5] we're not always diving into science there. It's sometimes comedy or commentary on science. I think people respond to it, again, because we're having fun with it.

Whether fun and comedy is the approach or something else entirely, you've got to pay attention to the storytelling basics. . . How do people see themselves in this story? The Jonathan Gottschall book, *Storytelling Animal*, talks about this idea that stories are ways for us to problem solve in low-stakes environments. It's how we make sense of the world. We've always got to keep our eye on this idea that we're trying to help people make sense of the world, but not in a way that we're looking down our nose at them, but in a way that we're helping them relate to it. If it's not relatable, then I don't think anyone cares.

KEIR: Absolutely.

DARYL: In all of my media experience, it was all about partnering. And it's something I don't see in Australia, museums genuinely partnering. I don't mean sponsorships. I mean saying, how can we extend our reach or how can we extend our impact or how can we tell the story differently, with a company or a group who is an expert in this space?

BRAD: I wanted to second your idea of partnership. But we've got to realise that we are publishers of real stories and real content that comes to bear in important ways. It's a resource that differentiates us from a lot of other corporations and organisations. I'm having some success starting to talk to partners who are interested in publishing with us, and I think that these partnerships are a way of getting outside of our own channels.

DARYL: And you have trust. And brand.

BRAD: Yeah, exactly.

DARYL: You bring such a significant value add to anyone.

BRAD: Yeah, yeah.

DARYL: The thing that you've got to work at is what you get back, because it has to be an equal fit. You both have to benefit or hurt equally. A good partnership has both parties getting something out of it.

We did a lovely small partnership with the *British Library* [in] celebration of the anniversary of the Magna Carta, which led to a collaboration between five of the key national institutions in Canberra. There is an original Magna Carta in Australia. It's on display in Parliament House and sadly not at our museum. So how do you compete with the original? We brought together seven different stories in five key institutions to create a rich and distributed experience. Our contribution was an Australian perspective on the Magna Carta, with a fully interactive experience looking at the key clauses that shaped Australian law. You could, literally, walk from *MoAD* to Parliament House, the High Court and the *National Museum*. It all added up to

---

5 The *Field Museum* has one of the largest and most complete Tyrannosaurus rex skeletons in the world, nicknamed *Sue the T.rex*. The Field has created a Twitter account for the skeleton which has over 50,000 followers and a wicked sense of humour. https://twitter.com/SUEtheTrex

something that we probably wouldn't have done had we not sat down and said, "Okay, who can we partner with? How do we make this a more valuable experience that is not just our small take on it?"

KEIR: I'm imagining a world where the polarisation of the citizenry is balanced by organisations like yours choosing to actively create partnerships with other organisations, other companies and other entities to create nuanced content, both digital and physical, that can act as a balance to the filter bubble-inspired, singular-looking, that we see now in our divisive politics. I would love a future where museums are constantly partnering to make new material and experiences. Because in partnership, you do things that you wouldn't do alone. A response to a society looking for material they're not finding elsewhere.

DARYL: Me, too.

BRAD: Well said. I wonder what role we can play in bringing communities together who otherwise wouldn't normally be connected to one another, and that we otherwise wouldn't connect to.

DARYL: One of the advantages to being in a small city like Canberra is the directors of all the cultural institutions have been getting together every couple of months. We have what's loosely called the culture club. We meet informally for lunch. We look at where we can partner, where are the shared issues, how we can assist each other.

It's transformed the impact that we have. So, I recommend creating a culture club for the museums in your city. Because a lot of the exciting ideas come out of informal conversations, where there is no real agenda. . . . I mean, I've jotted down five or six key ideas that have come out of this conversation that will influence our next strategic plan.

KEIR: Oh, that's fantastic! Thank you both.

The Center for the Future of Museums' *TrendsWatch 2019* identified the breakdown in trust, particularly as linked to truth and "fake news," as the first of five trends shaping museums' relationships with their communities (Merritt 2019). Within the U.S. and other countries, concerns about made-up news and information have risen, affecting public faith in government and commercial institutions, and even democracy itself. Such concerns are not evenly distributed throughout the population, but rather linked to political awareness. Highly political aware Americans describe increased recognition of made-up news than those who are less politically aware, while those with lower political awareness are more likely to be implicated in the spread of such news through sharing (Mitchell et al. 2019). This conversation with Brad and Daryl highlights some of the dynamics at play as museums seek to navigate this emergent information context, including public questions about museum neutrality, concerns about the links between sponsored and other forms of content, and how museums can create online and offline spaces for civic dialogue.

When social media rose to prominence in the mid-2000s, it was understood that it would lower the barriers to participation in public discourse so that anyone who wished to engage online could. Despite the ease with which these platforms can be used and accessed, at the end of the second decade of the 21st century, many people feel voiceless, unable to cut through the noise and competition to be heard. Brad provides insights into how his team approaches social media in a proactive way. Left unsaid is that museum social media managers are often junior and lowly paid, while being on the front line of all aspects of community engagement, including crisis management. Further, they are frequently called upon to manage negative public opinion when an organisation, it's leadership, curators or donors are under fire.

For more on trust, we suggest you read:

Conversation 1: Seph Rodney + Robert J. Stein
Conversation 2: LaToya Devezin + Barbara Makuati-Afitu
Conversation 4: Sarah Brin + Adriel Luis

For more on museum social media, we suggest you read:

Conversation 1: Seph Rodney + Robert J. Stein
Conversation 2: LaToya Devezin + Barbara Makuati-Afitu
Conversation 3: Lara Day + David Smith

References

"2017 Edelman Trust Barometer-Executive Summary." 2017. https://www.edelman.de/fileadmin/user_upload/Studien/2017_Edelman_Trust_Barometer_Executive_Summary.pdf.

"Democracy Index 2018: Me Too? Political Participation, Protest and Democracy." 2019. London. https://web.archive.org/web/20190630082129/**https://www.eiu.com/public/topical_report.aspx?campaignid=Democracy2018**.

Dilenschneider, Colleen. 2017. "People Trust Museums More than Newspapers. Here Is Why that Matters Right Now." Know Your Own Bone. April 2017. http://web.archive.org/web/20170815024834/**https://www.colleendilen.com/2017/04/26/people-trust-museums-more-than-newspapers-here-is-why-that-matters-right-now-data/**.

Dunn, Brad. 2017. "Look at the Entire Customer Journey 'From Couch to Couch'." CMS Wire. October 2017. http://web.archive.org/web/20171208130139/**https://www.cmswire.com/digital-experience/brad-dunn-look-at-the-entire-customer-journey-from-couch-to-couch/**

Mitchell, Amy, Jeffrey Gottfried, Sophia Fedeli, Galen Stocking, and Mason Walker. 2019. "Many Americans Say Made-Up News Is a Critical Problem that Needs To Be Fixed." Pew Research Center. 2019. https://web.archive.org/web/20190923155418/**https://www.journalism.org/2019/06/05/many-americans-say-made-up-news-is-a-critical-problem-that-needs-to-be-fixed/**.

Merritt, Elizabeth. 2019. "TrendsWatch 2019." Center for the Future of Museums, Washington, DC.

CONVERSATION 10

Arthur Cohen + Tonya Nelson

"If you look at it strategically . . . these organisations have the confidence and self-awareness to extend their role from that of expert to that of host."

– *Arthur Cohen*

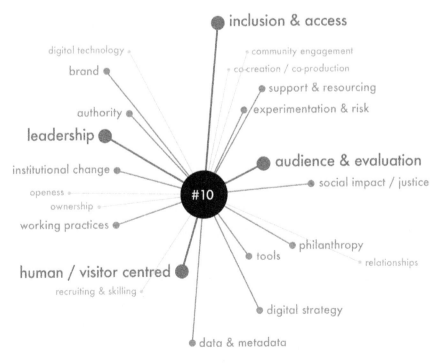

FIGURE 10.1

Tracking, mapping, and understanding the motivations and behaviours of audiences is critical work for museums. Two important publicly available tools for museums in the U.S. and the U.K. are *Culture Track*[1] and *Taking Part*.[2] The late 2017 iteration *Culture Track* coincided with the early 2018 *Culture is Digital*[3] report from U.K.'s Department for Digital, Culture, Media & Sport. When integrated with the *Digital Culture*, a study produced by Arts Council England and Nesta[4], these studies provide a detailed and nuanced view of digital cultural practices and audience attitudes and expectations in the U.K. and U.S.

Arthur Cohen (U.S.) is a leading strategic thinker and consultant in the cultural sector, especially for museums and galleries, and is the main spokesperson for *Culture Track* which is produced by the company he co-founded, LaPlaca Cohen. He is also a Lecturer at *Stanford University* in their inaugural Arts and Culture program in New York City.

As Head of Museums and Collections, Tonya Nelson (U.K.) oversaw the *Petrie Museum of Egyptian Archaeology, Grant Museum of Zoology, UCL Art Museum* and *UCL Pathology Museum*. Prior to entering the cultural sector, Tonya was a corporate lawyer and management consultant. Tonya was also one of the authors of the 2018 *Culture is Digital* report.

We brought Arthur and Tonya together to discuss shifting definitions of culture, changing audience expectations and motivations, points of access and inclusion, and new models of funding and philanthropy in the museum sector.

This conversation was recorded on June 1, 2018. At the time, Arthur Cohen was co-founder and CEO at LaPlaca Cohen in New York, U.S. and Tonya Nelson was Head of Museums and Collections at University College London, U.K.

KEIR: Arthur, how do you currently define culture and how have you seen that definition change over the last decade?

---

1 Culture Track is a longitudinal research study of the changing behaviors, attitudes and perceptions of cultural audiences in the U.S. that has been running every two to three years since 2001. https://culturetrack.com
2 Running since 2002, the annual *Taking Part* survey is a face-to-face household survey of adults and children aged five and over in England. The survey analyses cultural, digital and sporting engagement to provide a clear picture of why people do or do not engage. https://www.gov.uk/guidance/taking-part-survey
3 The *Culture is Digital* report sought to understand how technology can be used to drive audience engagement and how to boost the digital capability of cultural organisations in the U.K. (Dept. for Digital, Culture, Media & Sport 2018)
4 In 2017, the Arts Council England and Nesta released study of technology usage and impact among arts and culture organisations called *Digital Culture*. https://www.nesta.org.uk/report/digital-culture-2017

ARTHUR: I think anyone who attempts to identify or define culture with any specificity right now is playing with matches. The audience is generally ahead of the field. You get very different answers about what culture is when you start from the outside in than when you work from the inside out.

Every time we do *Culture Track*, which is the largest study in the world about the attitudes and behaviours of cultural audiences, when we shift the question from a somewhat traditional list of cultural activities – dance, theatre, art, opera, et cetera – and ask, "What is culture to you?" we get a much broader definition. Most significantly, that definition is now informed not by the activity but by the benefit that an activity triggers. Any definition of culture that is worth its salt at this moment is one that focuses on the impact that the experience has as opposed to the specificity of the experience.

KEIR: Is that something that you see as particular to this moment, or is that a trend that has emerged incrementally?

ARTHUR: We've done this study seven times since 2001. The direction has been consistent. The velocity has changed. Every time we've asked, "What is culture to you?" the responses keep expanding. Most significantly, at least in the U.S., we're in a new sociopolitical reality. The need for the things that culture is, perhaps, uniquely able to offer – stress relief, an ability to feel good and get away from anxiety-producing aspects that seem to be present in most people's daily lives – has increased significantly.

KEIR: How different are these definitions from the way culture is defined and described within the museum sector?

ARTHUR: A lot of the definition that arises internally is derived from how people were trained. If you think, for example, of a curator's training, their standards of success for what they do and the quality of the curatorial experience is not necessarily aligned with delivering an experience that resonates most with the audience. What I have found noteworthy and somewhat startling is the gap between the definitions of success when they are internally generated versus those that emanate more organically from the audience. Success from the audience's point of view has much more to do with their ability to connect with culture in an authentic and meaningful way – to feel it, and to find relevance to their lives.

KEIR: Tonya, what language are the people leading cultural institutions in the U.K. using?

TONYA: I think most people within the arts sector use "culture" to mean traditional art forms such as dance, theatre and museums. For the *Culture is Digital Report*, it was defined in those ways. But every other year, the U.K. does a survey called *Taking Part* where the government tries to map cultural engagement across the U.K. *Taking Part* has shown over a number of years that there are gaps in cultural participation between upper and lower classes and ethnic minority groups.

But other surveys and reports have shown that what people think of as "culture" is quite broad, and that most say that they are participating in culture. Because culture, if defined in terms like what you do on a day-to-day basis and how you engage with people, whether you get together with a group of friends on the weekends or the type of food that you eat, everybody can say, "I participate in culture." It's interesting to think about how the traditional cultural sector should respond.

KEIR: Does this evolution in how culture is defined by audiences change what museums think culture is? Are we seeing commensurate change in how organisations are framing their offerings or even the offerings themselves?

TONYA: I think cultural organisations in the U.K. are quite resistant to changing definitions of culture to respond to society's definition. To a certain extent, culture is tied up in this idea of expertise and skill. If you broaden the definition, that means you are diminishing some idea of expertise, skill, or gravitas and prestige. But in the context of digital, traditional cultural organisations have to start pushing themselves a bit more open to remain relevant. They have to think about the way people live and engage with culture in its broadest sense through digital technology and understand what that means for how people are going to engage with traditional arts and cultural forms.

ARTHUR: I think, generally, it's the same here. But these conflicts or pressure points aren't nationally or even organisationally defined. They exist within organisations themselves. The tension between curatorial and digital is a classic one, in part because they're looking at the same issues and analysing them in very different ways.

You're right that it often comes down to this false dichotomy that by somehow being sensitive and/or responsive to the audience, you are denigrating your expert authority. We don't believe that that's a trade-off. We believe that it's something that not only can coexist but can co-thrive. But a lot of people view it as threatening to do something that they consider to be ceding of authority in that dynamic between the cultural expert and the audience that it's ostensibly serving.

KEIR: My experience has been that the most successful projects that I've taken part in are the ones where there is an openness from everyone involved to move from "unquestioned authority" to "excited expertise," as that makes room for other voices and for other modes of dissemination.

When the goal to provide a meaningful experience supersedes the goal of, say, impressing your peers or getting published, there's room for an audience-centric approach that doesn't denigrate the curatorial or experiential intent. If the audience takes nothing from it, then that expression of culture is a failure even if you've been academically rigorous.

ARTHUR: I think that's true.

As consultants, pragmatism has to play a major role. We don't like to give advice or counsel that we don't think is going to be listened to or acted upon. For the conditions that you just spoke about to happen, it has to be clearly advocated from the top of the organisation. Unless the more traditional segments of the staff of the organisation know that the leader supports audience centricity, technological openness and all the concurrent dynamics around it, it won't happen. There needs to be leadership from the top. Sometimes that is initiated or charged by the dynamic between the board and the senior staff. Otherwise, what you're describing does not happen organically.

KEIR: I've heard it described as the executive sandwich. Where the board and the mid-level staff are on the same page and they need to squeeze the executive to agree to a different tone, a different voice . . .

ARTHUR: I don't agree with that. It makes a nice visual metaphor, but I don't see it working that way, at least in our culture. We have an entirely new generation of organisational leaders in this country, many coming from the country where Tonya's sitting. With that is a new generational perspective that is more grounded in the way the external world is changing and has a greater sensitivity to that. We're experiencing not just a hierarchical pressure, but a generational one, to say that the way the things have been done until the fairly recent past is not going to cut it in the future. This group of younger, more externally focused leaders will bring with them the sense of sensitivity to, and urgency around, addressing and incorporating these issues as a part of management practice.

TONYA: I agree. One thing that I appreciate about the U.K. is a body like Arts Council England, which sits between government, consumers of culture and cultural organisations. It has tried to oriented itself towards audience participation and engagement. With their funding power, they're pushing organisations to be more outward focused and to listen to their audiences.

The next-generation leaders are very outward focused. Currently, you still have people at the highest levels in the biggest arts organisations that are part of the old guard. They will say all the right things, but they're not creating the culture that's necessary for change. Certain positions within museums are valued more than others. For the most part, curators make it to the top in terms of leadership, and they tend to be the ones that are more focused on expertise. Those in public-facing jobs, like programming roles or visitor services, typically don't get to the top of the organisation. That reinforces certain things in terms of our outlook on audiences.

ARTHUR: In this country, that change, if it is being expressed or advocated for, is probably happening among the most enlightened foundations. The way that the grant applications are being rewritten to have a real impact metric associated with it and clear definition of success are forcing change that might not occur organically.

TONYA: Right.

KEIR: Arthur, from your experience with *Culture Track*, what are the major trends in how people are accessing culture, especially in institutions like museums?

ARTHUR: All the trends revolve around a couple of key dynamics. One is personal relevance. People won't participate in an organisation where they don't see people like themselves already engaged with and invited into. That doesn't necessarily mean that an African American won't attend a museum where they don't see African Americans, but that's part of it. It goes beyond that, too. They're saying that, "Unless the audience is, generally speaking, diverse and this is an institution that makes a conscious effort to open up to, invite in diverse, non-majority communities, then it's not a place for me." That's important in terms of getting people in, but also, in terms of the stickiness – whether this is an organisation that they want to continue to affiliate with and support.

That's when it goes back to another important dynamic, which is the idea that a cultural organisation, at least in this country at this moment in time, has to demonstrate its commitment to its community and social impact. It's the idea that, "I'm a member of this community. I'm not necessarily participating in your organisation because of my love of the object, although I might, but rather because I know or care about what you give back to the community and the impact that you have on the world within which the museum exists."

Both notions are super important. The idea of relevance, openness and invitation to start the process that can lead to an assessment of social impact shapes whether or not this will be a place that visitors want to continue to connect with and build a relationship with.

KEIR: Is that something that you witness in U.K. as well, Tonya?

TONYA: Yes. There are psychological mixed with geographic barriers that impede certain audiences from crossing the threshold of museum spaces. While those who frequently visit museums will have no problem leaving their neighbourhood to see an exhibition, there are those who find it difficult to travel more than a couple bocks from their homes. So not only do museums need to think about inclusivity in terms of the experience inside the museum, but they need to think about the experience of getting to the museum itself.

ARTHUR: That's right. Threshold fear is a real thing and it is supercharged by preexisting assumptions about inclusion, openness and whether this is a place that is for me. I always ask anyone in the cultural world to do the ongoing, informal field research that we do, which is any time you get into an Uber or Lyft and you're heading to a cultural organisation, ask the driver if they've been there. That will be a moment where you might hear an extremely clear example of how threshold fear plays out by whether that person feels that is a place for them. Whether they shrug it off or laugh it off. That moment indicates whether the organisation is effective at casting a broad net to indicate to everyone that they are truly focused on a diverse and broad level of engagement at all areas of expertise, and across all communities, or whether it's seen as serving a very small and narrow slice of the populace. These conditions are real.

KEIR: What you're describing seems to challenge the oft-cited "ladder of engagement," where a light-touch engagement with the museum, such as someone who comments on a social media post, is at one end and becoming a member or trustee is at the other.

ARTHUR: Are things like the ladder of engagement real? They're real if you're a part of the majority culture that doesn't have threshold fear, but that's a really narrow take on what encourages and motivates someone to participate in culture, particularly if you're concerned about expanding your audience. Any of these constructs that we take as being sacred truths, you need to ask, well, "For whom is that true? Is that really true for everyone?"

Last week, someone pointed out something really perceptive. In the States, we're dealing with fallout from the Starbucks incident in Philadelphia.[5] Are you familiar with that, Tonya?

TONYA: Yes, I am.

ARTHUR: Someone said, "Well, there goes the third space concept."[6] And they're exactly right, because . . . third space for whom? If the Starbucks model is about everyone's backyard and this need that we have to connect and interact with each other, but it turns out that you only get to play if you are someone who already feels like you're a part of that community, then you have to question the fundamental validity of the model. Does it actually work for an increasingly diverse audience that has different needs and expectations about what the cultural experience should deliver? We need to challenge all of our assumptions. All these models are so elegant that we take them for face value, but do we really think the ladder of engagement explains the relationship between the general audience and the cultural institutions today? I don't.

TONYA: I agree. It's incomplete in a lot of ways. One of the museums we have at UCL is the *Petrie Museum of Egyptian Archaeology*. We set on the path towards creating a new gallery in a collaborative fashion with the Egyptian community in London. We invited a group of people with Egyptian heritage to collaboratively design an exhibition space, which was all well and good.

---

5 In April 2018, two African America men, Rashon Nelson and Donte Robinson, were arrested in a Starbucks in Philadelphia, U.S., while they wanted for a third person to begin a business meeting. Starbucks staff had called the police. The event was filmed by another patron and went viral. Starbucks' CEO Kevin Johnson responded to public outcry regarding the arrests, calling them "reprehensible," and apologised to the men in-person. He ordered all stores closed for a day for mandatory training to tackle unconscious bias.

6 The concept of the third space (where home is the first and work is the second place) was popularised by Ray Oldenberg and other theorists where "your third place is where you relax in public, where you encounter familiar faces and make new acquaintances" (Oldenberg 1989) and include places such as cafes, social clubs, sporting clubs, public libraries and museums, parks and places of worship.

But when the Arab Spring happened, there was suddenly a big debate about the future of Egypt. Our group of collaborators got very passionate about using our space to have discussions about the social and political structure of the country. While the attention shifted away from an exhibition gallery, it became a space for them to gather and have discussions. They thought of the museum as a space they owned and could use in more personal ways. That's when the relationship moved from being transactional in nature to something deeper and more engaged.

The *South Bank Centre*, which is a multi-arts centre in London, has Brutalist architecture and at one time was a place where people thought, "I go to the theatre there and that's it." They did all sorts of things to make the space more open so that people would feel they could use it for anything, including church services and skateboarding. It has become a jointly owned and curated space – communities are producing their own culture there and then taking part with the culture that's being produced by the organisation itself.

KEIR: What is it about the South Bank and UCL that enables these audiences to trust them to host their cultural production? Are you trying to create trust? Is that a goal?

TONYA: In terms of the *Petrie* [Museum of Egyptian Archaeology], it was a turn of circumstances and an opportunity to say, "We hear our audience asking for something from us that we didn't envision, but we are open to and capable of providing the space for that use." I'm not as closely associated with the *South Bank Centre*, but they're not monitoring that space. The doors are open, and people can do as they like in that space, within reason.

Early on, it used to be that kids who were practicing break dancing or contemporary dance would use it as an open public space to practice. When nobody stopped them, they continued to do it. I think that happened with the skateboarding as well. When nobody stopped them, they're like, "Okay. Let's just continue to use it."

KEIR: That's excellent.

ARTHUR: There's a real important point here. If you look at it strategically, what's happening is these organisations have the confidence and self-awareness to extend their role from that of expert to that of host. A lot of organisations, most cultural organisations, don't have the confidence to do that. They view that as fraught with risk and, again, as ceding authority. But these examples, and there's a lot in the U.S. like these, show that the same organisation can play many different roles to its community and be activated in a number of different ways. It just has to open its mind up to that opportunity.

In almost any major cultural centre there're a lot of organisations that don't have the resources to have their own stage or gallery or their own gathering space. But that doesn't make them any less vital. So, the idea of the generosity can be displayed by a more significant or established cultural organisation giving them space to come in and invite their community. That's a very virtuous circle because, in addition to the short-term effect of getting over that threshold fear and having those people who perhaps have not participated in that organisational space before coming in, it helps in demystifying that space and signalling that sense of invitation. It increases the likelihood that those very people will return for other kinds of events and will form a different type of relationship from that which had existed before.

There's a lot of great things that can happen if an organisation has the confidence to expand the definition of its role and not only anchor it in this notion of authority.

KEIR: That's critical to me.

There are a couple of reports that came out of the U.K. in 2017 and 2018. The first one, *Digital Culture*, was a study produced by Arts Council England and Nesta, came out in 2017. And the second is the one that Tonya worked on which was a policy report called *Culture is Digital*.

Looking across these two reports, it's striking that in recent years, there's been a significant increase in the proportion of U.K. cultural organisations that believe digital is either important or essential to their business model. However, museums are doing fewer digital projects than they were four years ago.[7]

Could you explain the apparent gap Tonya?

TONYA: It's right for you to bring up that contradiction. I think that, as a general matter, the cultural sector is experiencing a digital project "hangover." Five or six years ago, Nesta, Arts Council England and the Arts & Humanities Research Council launched a major initiative, in which they allocated a large amount of money to digital R&D within the cultural sector. It felt like there were a lot of opportunities to experiment with digital, but the funding was project-based. People did interesting projects that were successful, not successful and somewhere in-between. And then it was, "Okay, what now?" The project mentality meant that organisations were going from project to project to project. That gets exhausting. They were not building expertise. It's like you're doing one-offs and then dropping them and then doing something else and looking for the next big thing.

I think the Nesta report revealed this phenomenon. People think digital is important, but they're doing less of it because they are tired of the project mentality around it. The DCMS [*Culture is Digital*, 2018] report ties the cultural sector into the creative industries where there is more of an infrastructure around R&D. The report presents a sense of optimism and excitement based on the idea that if we can bring the cultural sector together with the

---

7 Most critical here is the breakdown of digital activities reported by those surveyed for the 2017 *Digital Culture* (Arts Council England and Nesta, 2017) report, especially Figure 18 which visualises the "significant decline since 2013" of digital activities amongst digital leaders. Both reports show in multiple ways that digital is viewed as "critical" or "important" by the sector, although this sentiment is not evenly felt in different types of cultural organisations and in different geographies.

creative industries and the tech sector, there will be more opportunities for more sustained digital work. We could learn from the tech sector and the creative industries in terms of how they innovate, in terms of an iterative or prototyping approach to development, as opposed to a "big bang" project mentality.

KEIR: That's an important distinction. I noticed that the Knight Foundation, in its three years or so of funding museum technology, started with micro grants for projects. Within 18 months, they had pivoted to funding positions and putting people into organisations who could be digital change agents.

The two reports we just spoke about don't include much about cultural equity or diversity of programs, collections and staff, or have much space devoted to community co-creation and co-production. Are these topics part of the discussion in government? Is it just that they're not expressed in these reports?

TONYA: This is definitely a topic of discussion. We know that younger people want to be able to engage with culture on their own terms. They want a sense of ownership over cultural assets. Digital provides a platform for that. Where we need to move from, in the cultural sector, is using digital as a tool to disseminate what we've already developed in-house in terms of our bricks and mortar offer. It's like trying to translate that into a digital output as opposed to thinking, "How do we make our cultural assets more available and usable by our audiences so that they can be either co-creators or creators in their own right using the assets that we have?"

That probably didn't come out in the report, but that's definitely something that we saw. That the relationship between our audiences and the way that we use digital has to change in terms of thinking about digital as a platform for collaboration, self-production and/or creative entrepreneurship.

KEIR: In *Culture Track* there's the term "digital dilemma," which acknowledge that audiences are open to and expecting cultural institutions, heritage organisations and arts spaces to be engaging with them digitally. However, institutions aren't living up to the expectations of their audiences. Is that a fair summary?

ARTHUR: There's a couple more layers to that. The audience is ahead of the field in that everyone carries around this device, this smartphone, which is this remote control on their life and has an expectation of ease of use, ubiquity of access and applicability wherever they are. Cultural organisations are still figuring out whether that thing is the enemy. And if it's their friend, then what are they supposed to do with it? Are they supposed to create competing platforms?

The lessons about how to best use digital are not going to come from other cultural players. They're going to come from other fields because that's where best practice exists and there aren't examples within our industry of it. There're a couple green shoots, but nothing that we think is particularly compelling.

I think that digital has been the area, at least in this country, where the cultural industry has perhaps squandered the most money with the least bit of information behind it, because digital for a long time was seen as the goal as opposed to the channel. There was a rush to just go digital – get your app, do some flashy onsite screens and all that kind of stuff – all of which lasted about six months before it was replaced with the next wave of technology. But, cultural organisations don't have budgets that enable them to keep up with that. You wind up with a lot of outmoded stuff that no one used that came with promises that were never realised.

What's a cultural organisation to do? One of the codas in our practice [at LaPlaca Cohen], which I know not everyone shares, is that we don't believe that cultural organisations should ever be in the hardware business. The only hardware they should be concerned with is that which people are packing already, which is the smartphones in their pockets. They should rather be concerned with creating updatable learning platforms that respond to real needs and understand that digital is one channel through which they can activate their mission and vision. There are certain things that it can do uniquely well. Other things are probably best achieved through analogue and traditional channels.

The question should first be, what's the goal? What's the effect that you are trying to engender? From there, make sure that the mechanism through which you will deliver that goal is best aligned to help contribute to its success. Sometimes that answer will be digital, but sometimes it won't. I think that level of understanding is still emerging.

KEIR: One challenge is in how you think about funding digital projects. At *SFMOMA*, we try to spend about a third of our time, energy and money on content. We spend about a third on getting that content to the audience, so the UX and the software. And we spend about a third on making it operational and sustainable.

You might get money to do a project that is about disseminating content. If you spend all that money on the platform, then the content will suffer. If you spend all the money on content, but you don't deliver it to anyone, why are you creating it? And if you do something that can only last for six months and then disappears, or is attached to a fetishised technology or to a company that may go under, that may or may not share the same organisational values that you do, you end up in a situation the overall product might be deficient at launch and broken soon after.

In terms of the observation, that digital adoption should be led by the goals. In many cases, someone will come with an idea and I'll say, "I think what you're describing could be better served with a change in how we lay out the bollards and place signage so that people will know to turn left instead of right. An app isn't going to solve this wayfinding problem for you. Better signage will." I think that's healthy because you earn the trust of your peers if you don't try and turn everything into a technology.

What I'm excited about is people asking for advice on how to thoughtfully incorporate some digital tenets into an overall strategic vision rather than trying to break digital out or bolt it on. A digital strategy should only be an element of something that is actually about the overall mission and vision and a deep and rich understanding of audience.

ARTHUR: I agree with that. I had not heard the one third, one third, one third thing before, so I have to mull that over to see if it makes sense. But the second part is incontrovertible to me. The right question to ask is whether digital is going to be the most effective way to achieve your goal. The first part about the allocation of resources sounds a little bit like failing forward. The idea that everything you do, whether it is a success or not by initial terms, will make you smarter for the next time you do it. That is antithetical to museum practice as I've experienced it.

There seems to be a lot of risk associated with anything that can be labelled a failure even in the short-term. We hear a lot of resistance that is couched in terms like, "Oh, we tried that, and it didn't work." The idea that you would approach something with the notion that you might not achieve the goal you set out initially, but that it will move you closer to that goal ultimately and you have to take the long view, is both true and foreign to the cultural world that I've encountered.

KEIR: I'd like to challenge that. Actually, I'd like to reinforce it first and challenge it second.

ARTHUR: Okay.

KEIR: Being a part of the Silicon Valley tech world, at least geographically, we intentionally chose to create a lab environment in terms of practice. The Lab serves a number of purposes, one of which was to be able to have a peer-to-peer conversation with that community. I gave a talk at a museum technology conference, asking whether museum labs are a distraction. Someone in the audience stood up and said, "I lead an R&D lab and there is zero appetite for failure. We cannot fail, yet we use this language that is borrowed from the tech sector, but we're not internally allowed to have anything that isn't deemed a success." I thought that was interesting.

Something that *SFMOMA* – and SFMOMA Lab – has become known for in the last year is an SMS service called *Send Me SFMOMA* where you can text a word, colour or emoji, and we'll match it an artwork in collection and send it back to you in real-time. It received almost five million messages in its first year.

That project is built on a Collection API and used specific collection data, both of which were created for other purposes. Jay Mollica, creative technologist at *SFMOMA*, used credentials from a text chat bot project that had failed a few years earlier (something that never even went public), data that was entered into our collections management system years before the API existed and created a novel search interface which ended up producing something that was genuinely successful.

I don't think it would have happened if those tools weren't already in place through a series of things that were either modest successes or total failures. These small bets lead to something that is better, more than simply internal learning. We created projects that reached audiences. I see the point you're making, but there are at least some institutions where this is working.

ARTHUR: I think the story that you just described is a story of building internal credibility and permission to try. I would like to think that also includes permission to fail as long as the failure is coached in terms of incremental learning. I don't know how your department works, but I know a little bit about through talking to Chad [Coerver, Chief Content Officer at *SFMOMA*] and some other people that you work with, about how things happen internally. It seems that your team has been able to build up the credibility and the trust to enable people to go with you on the ride internally and take the chance on you. If that means a little bit of resource allocation to make that happen, you have a track record that has enabled you to do so.

I would like to think that track record also enables you to take some short-term hits as long as you can present them as learning that gets pushed forwards to make the next effort that much better.

I can't account for the person in the lab who got up and said, "We're not allowed to fail ever," other than to say I'm sure the cultural world is not unique in its resistance to failure and impatience, or unwillingness to understand the value of it. But I do think, at least at this moment, your experience is somewhat aberrant because of your success. I think you have a bit more runway than perhaps other tech departments within cultural organisations do both by virtue of the ecosystem that you're in, what part of the world you're in, as well as the savviness and success that your staff brings. I wish other organisations could model that in all of its complexity better than we see them doing.

KEIR: Okay.

TONYA: I agree with Arthur in the sense that, for the most part, the cultural sector does have a fear of failing and doesn't not offer space for failure. It is only in the large organisations that have decided that digital is a core part of their strategy where you see that kind of iterative experimentation that allows them to think, "Okay, we've tried one thing towards an ultimate goal and that didn't work out. Let's take what did work and add and build on that." That's the larger organisations and the ones that have decided that they're going to focus on digital innovation as part of their core strategy.

KEIR: Okay. That's helpful nuance and clarification.

I'd like to change tack a little bit and talk about funding and philanthropy, and explore how funding for museums and, more broadly, cultural organisations in Europe has changed since the global financial crisis and if that required institutions to rethink where the money comes from and, possibly, to be become more entrepreneurial. Tonya, when you talk to other museum directors about funding, what are their fears and what are their hopes regarding the money they need to sustain and grow?

TONYA: There is a general fear amongst cultural sector leaders in the U.K. around funding, as there is increased pressure from government to find alternative sources beyond government subsidy. There is pressure to think about how we can adopt U.S. models of endowments and more private philanthropy. There is pressure to think about new types of commercial income. Cultural organisations struggle with this from a skills point of view, but they also struggle with it from the more philosophical perspective of thinking that it is government's role fundamentally to fund culture and art.

KEIR: What are the practicalities of rethinking where money is going to come from?

TONYA: Over the 10 years that I've been in the sector here, there have been fluctuations in how organisations think about new business models. One interesting example came out of the *Culture is Digital* report was the idea of dynamic pricing for museums, where the price of an exhibition goes up at the peak times and goes down during non-peak times. This kills two birds with one stone: You can generate more money from exhibitions, while also making exhibitions more accessible for those people who can't afford to go to exhibitions. There's pressure to think about new ways of orienting the business, whether it's creating efficiencies using digital platforms or generating income from new forms of content.

There are short-term digitally driven solutions like crowdfunding. Then there are more long-term, culturally driven discussions around how much the wealthiest members of society should be contributing to the funding of the arts.

KEIR: In many ways, the U.S. model is evolving pretty quickly at the moment, as well. However, it's still the top 2% or 3% of American arts institutions that receive half of the available philanthropic dollars. Arthur, could there be an equivalent shift in how the American philanthropic-driven museum sector raises money?

ARTHUR: I think we have to be really specific and really careful about generalisations about market changes because there's very specific areas in which there's likely to be impact and other areas where that impact is easily overstated. That said, what we see as the fundamental shifts in the philanthropic models or philanthropic drivers in the U.S. include that the upper tiers of the philanthropy are going to be increasingly impact-driven and they will have increasing expectations of going into philanthropic relationships with cultural organisations with clearly defined sense of what success should look like.

We use the term SROI, Social Return on Investment, wherein there should be a clear expectation that if I give you X number of resources or dollars, I will see certain kinds of shifts, either quantitatively or qualitatively.

And I will be actively involved in the process in a relationship which is more dialogue-based than it has historically been in this country.

We certainly see more philanthropic sources being first and foremost concerned with community impact. It will be humbling but also a necessary adjustment in perspective for cultural organisations to understand that they're not competing against each other so much as they are competing with charitable organisations that can really make a case for the value proposition that results in social benefit. They need to understand and find better ways to communicate their ability to be competitive with such models, not that they can ever out humble a children's health organisation or other things that are clearly only about social impact. But there needs to be a sense that the articulation of the impact will be made more persuasively and clearly, because I don't think that has been a part of the game here up until now.

Look at it in terms of concentric circles. The circle of people who will support the cultural organisation primarily because of the cultural product it focuses on is a subset of the available population of potential philanthropists who will reward an organisation regardless of its particular focus if it can really articulate the ability to affect social change and contribute to a community.

I had a conversation with a billionaire philanthropist yesterday for a project we're working on in the Midwest. He's a board member of a cultural organisation. His response was consistent with response of many of his counterparts that we've spoken to, where we say, "Is the basis of your support because you are a collector and connect to this organisation by virtue of what it focuses on, or is it because you are an active community member and you primarily want to support the role this organisation plays in its community regardless of the specificity of its focus?" And he didn't even take a beat to say, "It's the second one, of course."

It always is the second one. The model in which a philanthropist is exclusively and/or primarily committed to an arts organisation because of the art is increasingly rare. It is decreasingly the driver of the affiliation and support that that philanthropist will exhibit.

KEIR: We covered a lot of ground and I have one last question before we go. Tonya, looking a generation ahead of where we are now. Do we have the cultural institutions here with us now that society will need in the future?

TONYA: I'm going to answer this optimistically. I think we have the raw material. We have the aspiration. There is a lot of good discussion going on about engaging the audiences of the future and making our institutions more inclusive, letting go of the authority and the narrow conception of expertise that stops people from being interested in our organisations. As we start to get to grips with digital, there will be new opportunities around inclusiveness as digital platforms enable new forms of collaboration and creative engagement. So, I'm going to say yes.

KEIR: Nice. And you, Arthur?

ARTHUR: I'm going to go with Tonya because I think, obviously, it's an impossible question because you're asking about potentiality and not current reality. I would like to think that we have the beta versions of future successful and relevant organisations in many of the ones that we have now, by virtue of a couple key assets that they have. One is the core things around which they are organised, whether objects, expertise or both. But the conversations are a lot more dynamic now than they have been.

And I think that as much resistance as we encounter in the field, it is still less than we have ever encountered in doing this for 25 years. There are more voices at the table than there ever have been. A broader range of expertise, a greater desire for relevance and to help people find and make meaning than I've ever encountered. I don't know that we have seen a lot of these organisations yet realise these potentials, but I would like to think that the ingredients are in place. And if we look at these as the betas for those future successes, yeah, let's go with yes.

KEIR: That is a fantastic, positive, upbeat way to end. Thank you both.

This conversation is shaped by Arthur and Tonya's broad and longitudinal understanding of how definitions of culture, especially in the digital and institutional realms, have changed in the U.S. and U.K. in recent years. Of particular importance to us is how museums can respond to what *Culture Track* calls the "digital dilemma" in the U.S., or as Tonya describes it, the digital project "hangover" in the U.K. These factors, as is argued throughout the book, can and should be responded to by a more digitally mature sector, one that adopts "an iterative or prototyping approach to development, as opposed to a 'big bang' project mentality."

Understanding if and how a museum is grounded in, and relevant to, its local community requires a strong grasp of who does and does not feel welcome. Information that many museums simply do not prioritise. The idea of "threshold fear" or whether or not the museum is seen as "a place for me" are real and genuine concerns. How a museum expresses *who it is for* goes well beyond what is exhibited, and who and what is collected. This expression is online, in social media, in email, in marketing materials, on street pole banners, in membership brochures and all the other words and images that museums send out into the world. It shows up in who is working at the front desk all the way through who is serving on the board.

Connecting these two ideas is a rich discussion of how institutional funding and philanthropy is changing in the U.S. and the U.K., which has implications for the other geographies covered in this book.

For a range of examples of digitally mature museum practice, we suggest you read:

Conversation 3: Lara Day + David Smith
Conversation 4: Sarah Brin + Adriel Luis

Conversation 5: Sarah Kenderdine + Merete Sanderhoff
Conversation 6: Kati Price + Loic Tallon
Conversation 7: Shelley Bernstein + Seb Chan
Conversation 9: Brad Dunn + Daryl Karp
Conversation 12: Daniel Glaser + Takashi Kudo

For more on threshold fear and other barriers to cultural attendance, we suggest you read:

Conversation 7: Shelley Bernstein + Seb Chan
Conversation 8: Kate Livingston + Andrew McIntyre
Conversation 11: Tony Butler + Lori Fogarty
Conversation 12: Daniel Glaser + Takashi Kudo

References

Department for Digital, Culture, Media & Sport, U.K. Government. 2018. "Culture Is Digital." https://www.gov.uk. March, 2018.
"Digital Culture 2017." 2017. London. https://www.artscouncil.org.uk/sites/default/files/download-file/Digital Culture 2017_0.pdf.
Oldenburg, Ray. 1989. *The Great Good Place: Cafés, Coffee Shops, Community Centers, Beauty Parlors, General Stores, Bars, Hangouts, and How They Get You through the Day*, 1st ed. New York: Paragon House.

CONVERSATION 11

Tony Butler + Lori Fogarty

"Being relevant to new and diverse audiences, and engaging with these audiences, goes way beyond marketing or audience development, or even education and community engagement programming. It gets to every aspect of the museum."

– *Lori Fogarty*

FIGURE 11.1

Both the Derby Museums Trust and *OMCA* are hyper-local organisations that act as exemplars of progressive, audience-centred museum practice. Tony Butler (U.K.) and Lori Fogarty (U.S.) and the organisations they led have used audience evaluation and understanding to inform their exhibition and public programs, as well as the supporting organisational structures. In different ways, they both seek to engage all audiences from those who are simply visiting the building through to those who participate in physical and conceptual co-production of museum content and experiences. They, and their respective teams, have placed social impact, wellness, trust and genuine relationship building with community at the heart of their very contemporary organisations.

Beyond their directorial roles, Tony and Lori contribute to pan-institutional bodies. Tony is the founder of the *Happy Museum Project* which explores how museums can contribute to a society in which well-being and environmental sustainability are principal values in order to create happy, resilient and sustainable people, places and planet. Lori is Director of the American Association of Art Museum Directors, an industry and advocacy association.

This conversation was recorded on May 18, 2018. At the time Tony Butler was Director of Derby Museums Trust, Derby, U.K. and Lori Fogarty was Director and CEO of the *Oakland Museum of California* (OMCA), Oakland, U.S.

KEIR: In many ways, the internet is the most global medium in the history of humanity, but one that paradoxically, in the West at least, can encourage some to seek out local or even hyper-local news, food and culture.

This idea of a local community seeking local culture seems foundational to both of your institutions. Tell me a bit about your museum, and its relationship to its local context.

TONY: The collection at Derby goes back to about the 1830s and the local philosophical society, when Derby was shaping itself around manufacturing and the building of industrial society. You can trace the whole civic place-shaping trend during the 19th century to cities like Derby and the institutions that grew up within industrialisation. The museum opened in 1880, and was part funded by a philanthropist called Michael Thomas Bass. He gave money to the local council, who decided a museum would be the best way to educate people and to provide a sense of civic pride.

The institution probably also exuded imperial power. I think that's writ large for lots of museums within England and Scotland that grew up during the 19th century. England was a very strong imperial power, so institutions like museums in regional cities became not just the expression of industrial and economic confidence, but also imperial confidence. Our collections are from all over the world. We have material from South Asia, from the Pacific

Islands, from Africa, North and South America, and areas where Britain had an empire.

The *Silk Mill* and *Pickford's House* opened in the 1970s, as industry began to decline or change, and culture and the local became an important part of post-industrial place-shaping.

Today, Derby is a multicultural community. 25% of the population self-identify as non-white British; 16% of population is Muslim. It's a city of a quarter of a million people, and there are 20 to 30 languages spoken within our schools. We are reappraising the purpose of the organisation as something that has a sense of local identity, but can also explore quite difficult notions of the post-imperial world for a generation of people who are growing up with no sense of empire. The generation of people in their 40s and younger have no resonance with the imperial nature of British society. There's a big need to appraise the collections in that context and explore multiple identities within our communities.

LORI: We're such babies compared to the British museums. The *Oakland Museum of California* opened in 1969 but came together from three museums that date back to the early 20th century. There was a small history museum, a natural history museum and art gallery that date back to 1910, 1916 and 1922.

When the institutions merged, the mission became focused on California in a deliberate attempt to distinguish and differentiate from the museums in San Francisco. Up until then, all three museums had collections and exhibition programs that were national and international.

It was seen as the museum of the people from the early stages. 1969 in the Bay Area was a time of heightened political activity and civil rights. The building was designed by architect Kevin Roach to be connected to the surrounding neighbourhood. The programming had a strong civic and education focus – not always realised. There's a famous story that the founding director, J.F. Holiday, a great California historian, was fired by the city commission that oversaw the museum six weeks before it opened because he created a community advisory council.

KEIR: Wow.

LORI: The city was extremely diverse, and there was a lot of outcry for the museum to be more representative of the community and to have community voices at the table. But the city leadership, the elected officials, city council, mayor, various city commissions, were all white, and did not advocate for Holiday's attempt to bring the community to the table. There was a tension from the very beginning. So that vision of being a museum of the people has been in our DNA for almost 50 years.

Oakland is very diverse, with large African American, Latino and Asian-American populations, and a growing immigrant community. A lot of people are moving into Oakland because of the cost of living in San Francisco. One of the biggest tensions in our city now is around gentrification and the potential loss of the histories, stories and culture of the longstanding communities – in particular the African American community – with an influx of new residents. That's something the museum thinks about a lot in our place as culture-keeper, culture-sharer. How do we welcome new residents and build our audience as the city changes, but also carry on the histories and traditions and cultures of Oakland?

We have definitely taken a more hyper-local approach in the last few years. This is another tension within our mission. We are the *Oakland Museum of California*. Our collections and the scope of our programming is California-wide. There's always a discussion within the museum with our board and staff about, "Should we be reaching out to communities in California beyond Oakland and the East Bay?" But we have a lot of work to do to build our audience in Oakland. There are people within a five- or 10-mile radius of the museum who don't know about us, and we are the largest cultural institution in Oakland. Our foremost priority is building that local audience.

KEIR: You've used the word "community" a number of times. Lori, at the *OMCA*, what do you mean when you say community?

LORI: We mean the natural audience of the museum where transportation or travel shouldn't be a barrier. That audience is also among the most culturally diverse within Oakland, including low-income neighbourhoods. In these last few years, through an initiative funded by the James Irvine Foundation, we have explicitly focused on the neighbourhoods right around the museum, including West Oakland, which is a largely African American community; Chinatown; East Oakland, which is highly Latino; and the uptown neighbourhood, which is a very mixed community, [with] a lot of newcomers.

We are striving to grow our culturally and economically diverse audiences to be reflective of the broader population in Oakland. It's harder for us to measure what the economic diversity is, because a large part of our audience won't provide information on income. We also think about other areas for audience growth and incorporate that in "community" as well. We have focused on growing our audience of families with young children, youth and young adults.

KEIR: I remember you talking a few years ago about a neighbourhood study of volunteers and audiences. You discovered the ZIP codes around the museum weren't strongly represented in your attendance. This feels like a concrete response to that. Is that fair?

LORI: Oh absolutely. We target neighbourhoods for on-site and off-site participation. Some ways of engaging with these audiences are to be out in the community and at community events. We also looked at what these neighbourhoods might hope to gain from the museum, like green space.

KEIR: Has this changed who is coming to your museum? Who works for your museum? How you program what's in the museum?

LORI: It's changed everything. It's changed how we think about our composition and skill sets for our board. Our staff. Our programming. Being relevant to new and diverse audiences and engaging with these audiences goes way beyond marketing or audience development or even education and community engagement programming. It gets to every aspect of the museum.

KEIR: Tony, in the *Derby Museum*'s 2018–2022 business plan, you state, "In 2020, the Silk Mill will reopen as a museum of making, the first of its kind to be *entirely co-produced with the public*." What does that mean?

TONY: It means giving people a concrete role in a range of aspects of the museum. People have been involved with the architects to help design spaces. They've helped select, conserve, restore objects. They've helped with the decant and recant. They've become part of the decision-making process about what the future of the institution can be.

There's a practical element, too. In the prototype phase in the *Museum of Making*, we built a workshop. It has laser cutters, 3D printers, routers and woodworking facilities. People could sign up for a two-month program, become skilled into using this equipment, and design and make cases, fixtures and fittings for the new museum. That's effected our other sites, so members of the public are actually making fixtures and fittings for our new natural history gallery and world culture's gallery *in situ*.

We ran a give-and-get volunteering program using our workshop. For instance, we have a guy that would use our equipment to make bespoke skateboards. In return, he would give a day per week training schoolchildren how to code.

This notion of a museum built on relationships rather than management structures or the traditional transactional structures was really important to this project. The change for us has been physically enabling people to make the museum, be it through design, through making cases, through the selection of objects and stories and narratives. They've had influence in decision-making at board-level, they've been able to contribute, right at the beginning of the process, to try to define what the museum could and should be. That's what we mean by co-producing.

We take all this to the historical antecedent of the city. The *Silk Mill* was built in 1720. There's 300 years of making at that site. Today we have manufacturers like Rolls Royce and Bombardier in the city. They have workers that want to contribute to the development of the new museum. Rolls Royce are working with their apprenticeship schemes and their STEM [Science Technology Engineering Math] ambassadors to train and to develop people connected with the museum.

We begin construction on the new museum in a couple of months, and it's due to open in 2020. The museum is a place of its history, but it's also useful in providing people with training, opportunities, skills [and] a meeting place. It really does inspire people into the future.

KEIR: There are very few institutions that are giving their audience this sort of agency. I'm curious how that's been received within the museum?

TONY: When I was running the *Museum of East Anglian Life*, we repositioned the whole organisation as a social enterprise, using all the assets to develop community capacity and social capital, and to be useful. We ran training programs, within which there was an element of co-production, as graduates would become teachers for the next group, and those skills were passed on and on. You would find an antecedent in any rural community going back centuries.

Then I started a program called the *Happy Museum Project*, that set up a community of practice in U.K. museums where well-being and environmental sustainability was seen as two sides of the same coin. We encourage museums to experiment with participatory practice.

When I came to Derby, it was the question of upscaling to an urban and city environment. There were extraneous circumstances that helped. *Derby Museums* used to be part of the local council. It became an independent charity just before I arrived, so there was a more independent management structure and a board of trustees. We had austerity policies come in. Much of the state funding had reduced, and we had to restructure. That provided the impetus to embed a different set of values and practices.

One or two old school curators decided that it wasn't the kind of institution they wanted to work in, but with a new generation of curators, this is becoming embedded practice. Our fine art collection consists of the best collection of work by Joseph Wright of Derby in the world. It's a designated collection of outstanding national significance. Our art curator is as much interested in setting up project labs and community participation as our social history curator. There can be degrees of co-production for different types of work. Some of our work is incredibly participatory and genuinely co-produced, and other parts might just have a little bit of participation. The organisation is flexible enough to have deep engagement in certain areas and light engagement in others. But the time of one-way dialogue and a relationship built on transaction between the institution and the visitor, that's over.

KEIR: Let's hope so.

When you posted the business plan on your website, you also Tweeted, "Hold us to it in four years' time."[1] I'm holding you to it a bit earlier than that. It says, "By 2020, 100% of our collections of making will be accessible at the Museum of Making at the Derby Silk Mill." (Derby Museums 2018) What do you mean by "accessible?" Does everyone have a key to the

---

1 In March of 2018, Tony Butler Tweeted: (https://twitter.com/tonybutler1/status/978996298232758273): "We've just published our Plan for 2018-22 @derbymuseums Download it here . . . Hold us to it, 4 years time!"

warehouse? Do you mean you'll digitise the entire collection and put it online? Or that you'll hang the whole thing salon style?

TONY: A bit of all, but mostly hanging everything. We have four floors in the *Museum of Making*. Each will have a graded level of interpretation, from iconic objects to more traditional narrative, to dense open-display storage. With around 70,000–80,000 objects, it's a varied collection, and you will find the majority of those objects on display. Our ephemera, like the photographic collection, will be digitised and available at the site and online, but the intention is that as much of that collection as we can will be accessible and physically available to see.

KEIR: That's very interesting.

Lori, how has access to funding and resources affected your organisation, opportunities and your strategies?

LORI: It's what keeps me up at night.

Until 2011, the museum was a department of the City of Oakland. About half of our staff were city employees. I was a city employee. However, city funding was not general funding. It funded specific positions, like custodians and security guards, but also curators, registrars and educators. It covered core collections, and the care and upkeep of the building through our public works department, but it didn't fund programming, marketing or exhibitions.

Funding was always a challenge, especially right after the economic downturn. The city was in financial crisis. We made a difficult decision to transition from being a city department to independent operation. We were able to negotiate with the mayor and the city council, so now we operate within a 10-year grant agreement from the city. The challenge is that that city support is declining from about 45% of our budget to closer to 20% over this 10-year period. We've had to be very aggressive in building revenue in all other ways. We're in the midst of a major comprehensive campaign to build operating support as well as endowment and investment support, and growing our earned revenue.

This is our great challenge right now. We talk all the time about our goalposts, picturing an American football goalpost with two prongs. One of the prongs is financial sustainability and the other is social impact. We are in the early processes of being able to more clearly articulate what we mean by social impact and how we measure that. But trying to find that right balance between being accessible and engaging new audiences and underserved communities, and growing revenue as well as philanthropy is a careful balancing act.

We're not there yet. We've had some success in growing our revenue, but our financial sustainability is still a real challenge. After 2021, we don't have any guarantee for ongoing city funding. We certainly plan to renegotiate, and hope for a slight increase. But the City of Oakland is so stretched, with its needs around homelessness and public safety, et cetera, it'll be an uphill climb.

KEIR: That's a pretty steep decline. I imagine that sense of uncertainty impacts your ability to attract long-term corporate funding, if you don't have the same commitment from the city that you can lean on and leverage.

Tony, what's the funding mix for the *Derby Museums*, and how are you supporting the *Museum of Making* project?

TONY: We also used to be part of the local authority, until 2012, when we became an independent trust. We are a charity, but the collections and buildings are owned by Derby City Council, and maintenance of those assets is still in their hands.

LORI: Our collection and building are also city-owned.

TONY: It's almost the same model, isn't it?

LORI: Very similar.

TONY: When I took over in 2014, 97% of our income came from two sources: the city council and the arts council. Arts Council England is a non-governmental public body that funds many aspects of arts and culture, including arts organisations and some museums. Something like 65% of our funding was coming from the local council and 32% was from the arts council.

Since 2014, our public funding has reduced by 40%. We've had to make serious efficiency savings, and to think clearly about generating earned income. Our earned income now is 27% of our income. Some of that has been linked to our public sector income going down, but we've also begun to think more clearly about how we set up our assets. We now have a well-established venue hire program, we're more commercial around our shops and cafes, and we have an assertive strategy for on-site donations. However, we don't charge admission within our museums, they're all free.

We have odd laws in Britain around donations. We have a thing called gift aid, which means that you, a taxpayer, can claim an additional 20% of the donation that you give on your taxes. There's a new regime coming called Exhibition Tax Relief, which means you can claim back all the tax on the building of new exhibitions. So, there's a range of schemes that support the organisation.

We're developing an endowment, which we haven't previously done. We have a target of raising two million pounds by 2021. We've raised a quarter of that in about 18 months. We have a patron who's the Duke of Devonshire, who's very influential, and we've been able to leverage in his influence to begin to find high-net-worth individuals. But there's no guarantee that we're going to get any public funding beyond a small amount after 2020.

In terms of capital development, the heritage lottery fund is the principal funder for the new *Museum of Making*. That's a £17.5 million project, and £9.5 million of that is coming from the Heritage Lottery Fund. We've got another £2.5 million from the Arts Council Lottery Fund, and the rest has come from a central government fund designed to stimulate the economy in post-industrial areas. Then we've had money from the Trust and Foundation. The struggle as ever is going to be revenues. But there is, I think, an understanding amongst city museums in the U.K. that we'll end up developing along the lines of the American model, which is a mixture of public funding, income through endowments, donations and fundraising, commercial sponsorship and commercial activity.

KEIR: It's incredible how seldom people draw the line between the global financial crisis and the remaking of the cultural funding landscape in Europe, the U.S. and other parts of the world. So many directors I've spoken to outside the U.S. have gone from being 70, 80, 90% government funded around 2009, all the way down to receiving 10 or 15% of their funding. It's turned museum directors into entrepreneurs, because you can't be relevant to your community if you're not financially sustainable.

LORI: That's right.

KEIR: Australia and New Zealand largely skipped the GFC, so they haven't quite had that reckoning yet.

I think it's an interesting space for the sector to be, where we want to provide high social-impact programs, but also need to keep the doors open.

TONY: Some of our new donors are donating precisely because they had a formative experience in the museum –

LORI: Right.

TONY: And they can see the power of what the museum has done for them, or their families, and want to give something back. That's genuine philanthropy. There's a statue in Museum Square, just outside our museum, of the philanthropist who gave the money for *Derby Museums* back in 1880. It's almost as though we've come full circle in 140 years.

LORI: Where we are seeing some success in attracting new donors is because of the *Oakland Museum*'s commitment to community engagement, to social impact, to being truly welcoming to the broadest audience. We now have the metrics to show the change of the composition of our audience and the public response to our programming. People inside the museum often see financial sustainability and the need to generate revenue as in opposition to our commitment to social impact. In fact, they're mutually dependent and completely integrated.

So, I think it is a shift and it's changed our jobs a lot. The director role has long been a fundraising role, at least in the United States, but now it's thinking in new ways about business development. We created a new position this year, Director of Business Development, to think about ways to generate revenue and to leverage our assets.

TONY: Absolutely. I've been more influenced by the idea of asset-based community development than any other form of thinking. About how a community can use its social, cultural and physical assets for broader benefit. Once I started to think about the museum in those terms, it was liberating.

KEIR: Looking at the *Oakland Museum of California* org chart, it reads like a flower. I don't know if that's deliberate. Could you describe it, and tell me the benefit of having an organisation structured around centres, rather than a hierarchy?

LORI: Our org chart used to look like a flower, like petals around a centre. We've since consolidated into four centres, so now we're more like a four-leaf clover. This happened when we transitioned from the city. One of the incredible benefits of that was the opportunity to completely rethink the organisational structure. We had job descriptions that were literally 40 years old. The city laid off all the city employees, and our non-profit entity was hiring them back. We decided to rewrite the job description of every person within the museum, and completely restructure the way that we were organised. 80% of the staff had to reapply. It was as much culture change as structural change. We were reinventing a 40-year-old institution. We reinstalled our 100,000 square feet of gallery space in project teams that grew from multiple centres, and with multiple areas at that point, multiple departments. That's the basis for our thinking about the centres.

I think there are two that are unique. One is the centre for experience, development and collections, which includes the curators and collections staff, but also what we call "experience developers." They are the staff often called exhibit developers, who are really the visitor advocates, working with the curators to essentially do what Tony described earlier, although maybe not quite as ambitious a scale of co-production. But that is where community collaboration comes into our development of exhibitions and projects. The experience developer, along with the curator, leads that process. In that centre, curators, conservators and other collections staff work with experience developers and visitor evaluation, always having the visitor voice at the table.

The other centre that I think is fairly unique is the centre for audience and civic engagement. In this centre we have learning initiatives, which is museum education, primarily K-12, and docents, with public engagement, marketing and communications. Engagement and marketing have to go hand in hand. So much of marketing and communications is about the social experience and the programmatic elements. We're also thinking about that continuum of how we reach the visitor from the moment we connect with them, through the on-site visitor experience, into deeper engagement through education and public programming.

We still have somewhat of a hierarchal structure in reporting relationships. The heads of each of those centres report to me, along with the director of HR. But we work a lot in cross-functional project teams that draw from all levels of the organisation and across centres. The blurring of hierarchal boundaries takes place through that project work.

KEIR: One of the things I've seen consistently when working with Bay Area technology companies is an ad-hoc, project-based approach to most new development, whether it's product, program or experience, with a constellation of people that have the right skillsets – which sounds like what's described in the a human-centred design toolkit you published Tony.

TONY: Ah, yes.

KEIR: There are workflow diagrams, which look very similar to design thinking and agile software development workflows. It's interesting to me how these practices are extending beyond the development of digital technologies into audience-based product development. Where did the toolkit come from?

TONY: The human-centred design toolkit is the brainchild of Hannah Fox. Hannah is the *Silk Mill* project director, and about to become our Director of Projects and Programs. Prior to coming to work in Derby, she helped run a design firm and was also developing heritage and arts projects at the community level within the city. There was a national program called Creative Partnerships in the early 2000s that she ran. She entered the organisation about 12 months before I got here, and bears credit for embedding human-centred design thinking and the nascent stages of our co-production ethos into the rest of the organisation.

We wanted this work to be open source, and we've had four or five iterations of the handbook on our website. And you're absolutely right. It takes the principles of design thinking from product design and from the realm of tech and overlays in community work and the making of a museum. It's not just applicable to exhibit design or project design, but also how we work with our communities. The design handbook also uses our principle collection, the work of Joseph Wright of Derby, to symbolise each step of the design development phase.

We think it's been a really positive way to embed it within the organisation. In terms of staff, we're a bit smaller than Oakland, and we've gradually become flatter as an organisation mostly through efficiency savings. The salary differentials between me and the chief executive and the next layer of management staff is fairly small, and throughout the organisation the salary differentials are fairly small, which means that it's easier to develop a more flexible and eccentric way of working. I suspect in reality we're working in a fairly similar way to Oakland. We don't have a collections team, we don't have a curatorial team. We have curators – they work alongside our participation staff and our learning and development staff on projects.

When we start a project, everybody's involved, and then it narrows down to smaller project teams.

KEIR: Has your mission changed? Are you still a museum?

TONY: We are. I think this is a result, although we wouldn't explicitly say it, of a policy of greater social purpose within cultural organisations. It goes back 25 years or so where learning and participation become absolutely central to the ethos and the mission of museums in the U.K. From the new Labour government of 1997 onwards, funding, support and policy was far more linked to social outcomes and being useful then it was to general preservation of cultural heritage.

The U.K. is the only country in the world where arts organisations have to have an environmental policy and have to commit to reducing their carbon footprint, if they don't – they won't get funded.

KEIR: Lori, when I first moved to the Bay Area, *SFMOMA* had just closed. I remember visiting the *Oakland Museum of California* to try out some ideas and prototypes. Your staff were really generous, allowing us to prototype inside your galleries and with your visitors.

I remember coming in early to explore the museum before one of the first meetings, and I found a couple of very simple technologies. One was a sticker. There was a big map of the world and I could take a little red sticker and stick it on Sydney, Australia and show where I was from, and see where everyone else was from. I felt acknowledged in a simple and quick way.

The other one was a digital self-portrait drawing application. I think it's called *You Are Here*. It's a drawing activity integrated into the galleries. I was helped through a series of steps to draw a self-portrait. It gave me some tools and ways to think about portraiture and self-portraiture. Then my portrait joined other digital portraits that were digitally hung within a constellation of all manner of painted portraits from your collection. Both projects would have cost much less than I usually spend on *SFMOMA* technology initiatives, but both felt inherently democratic and welcoming.

We were talking about putting the audience at the centre of the experience . . . is that to move from visitation to belonging?

LORI: Oh absolutely.

KEIR: Is belonging a word that Oakland uses? How do you quantify belonging?

LORI: Both of those activities are powerful examples of what we're trying to do. Your earlier questions about, have we changed our mission, have we changed how we measure our work? Are we still a museum? Those are central to our work now, and how we think about our impact and our mission.

We did change our mission statement in 2013. Now our mission statement is "to inspire all Californians to create a more vibrant future for themselves and their community." It is a shift to say we're partnering with communities, and it's really about well-being.

We have worked over the last couple of years on this. Museums in the U.K. have been far ahead in thinking about well-being. We think about how to articulate and identify what our social impact is and how to measure it.

What we are striving to do is address social fragmentation and to build stronger social cohesion. That is something we believe our museum can fairly uniquely do in Oakland, with the enormous diversity of audiences. In our Friday night events and exhibitions, like the Hip Hop exhibition or the Black Panthers exhibition, it really is about bringing people of all different backgrounds together with this shared experience of welcoming and belonging.

We're identifying the outcomes for this social impact as visitors telling us, "I feel a sense of welcome and belonging." So yes, the word very explicitly used. "I see my own story." And, "I see the stories of other people that I can identify with and understand, and I connect with my neighbours." Those are three or four outcomes that we're looking at. We're going to be starting this year how we build in some way to measure those through our on-going visitor evaluations. We're not going to ask people directly, "Do you feel like you belong?" But we're trying to get to some of those measurements.

Even in our existing audience evaluation when we have the open-ended questions, particularly around evaluation of exhibitions, people do use those words. Or when we've done evaluations of our Friday nights[2], the feedback that we've received from our visitors very much deals with those kinds of outcomes, "it's empowering, I see myself, I feel like I belong here, I see my story told, I see surprising stories of other people that I never knew, that I now connect with in new ways."

TONY: I've been following Oakland's approach through your online publishing over the last year or so. I'm really interested in the way that you're beginning to capture that sense of community engagement, community capital. When I set up the *Happy Museum Project*, it was funded by an organisation called the New Economics Foundation. They produced a program called "The Five Ways to Well Being" (Aked and Thompson 2011). These are five very simple ways in which you can begin to measure the well-being impacts from organisations and how people have benefited from them: connect, keep learning, take notice, be active, give back.

Over time there's been a range of measurements and ways to articulate that organisations are experimenting with. They can be really simple, like one museum we developed with the *Happy Museum* called the *Mood Tree*, which had two trees, before and after, and a series of coloured leaves. It was green if you were happy and red if you were sad, and amber halfway through. You would put on the tree how you were feeling before and after an activity, with hopefully a different colour after that activity.

We worked with the university and we took cortisol cheek swabs before and after a making activity.[3] The results showed that being engaged in an active and social activity meant that cortisol levels went down, which means the immune system is strengthened and it's good for well-being.

At the *Museum of East Anglian Life*, we used a system called the Social Return On Investment model, which quantifies social outcomes in an econometric sense. There's been a big study using SROI in Manchester. Esme Ward, who's now Director [of Manchester Museum], but was then the Head of Learning, did a big project around volunteering, and measured the social benefits of volunteering using econometrics (Lackoi et al. 2016). For every pound invested in an activity, you got four or five pounds of social value. That's measured by whether people have stronger relationships with

---

2 Every Friday night, from 5 to 10pm, the *OMCA* hosts a popular street party inside and outside their museum with music, food trucks, beer and wine, kids activities, local makers selling their handcrafted pieces and after-hours access to the museum.
3 In one such experiment, through measuring cortisol levels in saliva, the *Happy Museum Project* showed stress levels in London city workers significantly decreasing after half an hour in an art gallery.

their families, whether they're going to the doctor less, whether they're living more independently, if they're unemployed and they're going to get jobs. It's a whole range of proxy measures to measure well-being based on the investment given.

The *Museum of Making* project is going to be evaluated using a similar social return on investment model. We're looking at how engagement in the museum and involvement in building the museum impacts on the sense of belonging, whether they know their neighbours better, whether they feel physically fitter.

I think you're seeing a lot of different ways emerging to measure well-being impacts in a population. Over time, those studies will be more useful. Certainly, we've found that some of our supporters have given us money on the basis of some quite hard evidence about the well-being impacts of our work.

KEIR: At *SFMOMA* we, led by evaluator Kate Livingston [also featured in this book], added two questions into the post-visit evaluations of our digital interactives. The first question was, "Do you think *SFMOMA* cares about your experience?" And the second was, "Do you think modern and contemporary art is relevant in your life?" Then we cross-tabulated that against a baseline of people who had and hadn't experienced one of these interactives.

There was a good but modest movement forward, like 15%-ish improvement for people who witness one of the interactive, interpretative, story-led experiences within the museum. However, people who did something with their hands – touched the button, turned the dial, swiped, read, pressed play or stop, generated an image, created something through an activity – it was a 45% jump in terms of, "Do we care about your experience, and is this relevant in your lives?" Taking everything else out, simply whether you witnessed an interactive engagement or you interacted yourself moved the needle.

TONY: Fascinating.

KEIR: The 2017 *Culture Track* report listed "Have fun" and "Feel less stressed" as two of the top motivators for attending a culture experience in the U.S. (LaPlaca Cohen 2017). How important to your organisations are those values? What things that you're doing or experimenting with now do you think are going to be just commonplace ten years from now?

LORI: We think a lot about both of those categories. Like other museums we do an NPS [Net Promoter Score] as a proxy for visitor satisfaction, and we rate quite high on that. When we did some visitor research a couple of years ago trying to get to what was behind our NPS, the number one thing people said was "that we have fun here." Sometimes there's mixed reaction to that internally, because you think about how much effort we put into the learning and the content. But, as my colleague Kelly McKinley said, most museums would give their eye teeth to have their visitors say that they're fun.

TONY: Yes.

LORI: It's interesting that we haven't put that as more of a specific outcome with social impact, but we do think about people having a social experience, a chance to be with family. At certain moments in our community, like after the November election in 2016, or after the Ghost Ship fire, a very tragic incident that happened in Oakland with a loss of life of artists and creative workers in a warehouse fire, visitors say things like, "I came here to have my faith in humanity restored." Or, "I came here to be with other people, to share the space." There is need in these times to have a place where the community feels like lots of different people are coming together, and I can bring my family, and I feel good about being in a public space. Those are definitely factors that we think about.

It is shocking to me how few art museums in the United States do any kind of real visitor evaluation. As recently as a couple of years ago, at the Association of Art Museum Directors, we did a panel around visitor evaluation and asked that room how many museums have evaluators on staff. I think two museum directors raised their hands, other than those of us who were on the panel. That will become more commonplace.

The other one is thinking in new ways about creating social experience. We were talking here about interactive and hands-on elements that are not relegated to outside the gallery space. I think that will change, and there'll be more within the exhibition spaces. Where we're trying to experiment now is bringing live programming within gallery spaces themselves, and I think that there won't be such a clear line between the sanctity and quiet of the exhibition space, and that all of the social, interactive, fun elements. That there'll be more of a blurring of those lines.

KEIR: That's fantastic, thank you, Lori. Tony?

TONY: I would agree with that. The only thing to add would be, I think there will be less of an expectation to see things finished. I think galleries will never be finished in the future.

LORI: Mm-hmm, mm-hmm (affirmative).

TONY: People will see that they have a right to participate and galleries will change all the time, so new things will appear, things will be taken away. In terms of the exhibit design, I suspect that you won't have beginnings, middles and ends. You're just going to have beginning and forever middles. And it might be that the production values are a bit lower, and things are cheaper. In-built obsolescence won't be a feature of the public space anymore. That's certainly a thing we've found with the natural history gallery and the world culture gallery we're building up – people expect things to change all the time.

Secondly, the notion of understanding what cultural capital means, and why people don't come to museums, will be far better understood. At the moment in Britain, non-attendance is still seen as something that is predominantly economic, rather than cultural. With the research that we've done in areas of low participation, the perception of the museum is too much like school, and as purely a place to learn, and people have bad experiences at school and don't want to come to a museum. That museums and galleries aren't fun, and they're places that are not for the likes of us. We'll be far more understanding and willing to challenge these deep-rooted deficits in cultural capital within our communities, and not put it down just to economics and social barriers.

LORI: I completely agree with Tony's points. Trying to move away from permanent collection galleries or permanent exhibits to building, prototyping and iterating. These are things that science and history museums have done historically better than art museums, but we're moving in that direction.

TONY: The museum is everywhere, is what we're trying to develop an idea upon.

KEIR: Indeed. Thank you both.

The museums and centers that Tony and Lori oversee are engaged in a very contemporary – but also timeless – project that takes a local, or even hyper-local, approach to community access and engagement and community co-creation and co-production that extends right through to their exhibition-making. This commitment defines their place and relevance within their community that goes beyond numerical inclusion to something that looks a lot more like community ownership.

This conversation explores social impact, through an asset-based community development lens, in a concrete, exemplary way building on the high-level discussion between Tonya and Arthur and, in doing so, offers some new ways to think about "who museums are for." Both organisations provide democratic, welcoming, participatory technological interventions, both in the galleries and behind the scenes, that serve to make people feel seen and valued, leading to a sense of belonging – while still creating spaces that are fun to visit.

For more on community-oriented practices, we suggest you read:

Conversation 2: LaToya Devezin + Barbara Makuati-Afitu
Conversation 4: Sarah Brin + Adriel Luis

For more on comparative institutional structures and iterative working practices, we suggest you read:

Conversation 3: Lara Day + David Smith
Conversation 5: Sarah Kenderdine + Merete Sanderhoff
Conversation 6: Kati Price + Loic Tallon
Conversation 7: Shelley Bernstein + Seb Chan
Conversation 9: Brad Dunn + Daryl Karp
Conversation 12: Daniel Glaser + Takashi Kudo

References

Aked, Jody, and Sam Thompson. 2011. "Five Ways to Well-Being: New Applications, New Ways of Thinking." New Economics Foundation. http://web.archive.org/web/20190620111351/**https://neweconomics.org/2011/07/five-ways-well-new-applications-new-ways-thinking**

Derby Museums. 2018. "Derby Museums Business Plan." Derbymuseums.org. https://www.derbymuseums.org/wp-content/uploads/2018/03/0383_DM-Business-Plan_A5-12pp_digital.pdf.

Lackoi, K., Patsou, M., and Chatterjee, H.J. et al. 2016. "Museums for Health and Wellbeing. A Preliminary Report, National Alliance for Museums, Health and Wellbeing." https://museumsandwellbeingalliance.files.wordpress.com/2015/07/museums-for-health-and-wellbeing.pdf.

LaPlaca Cohen. 2017. "Culture Track 2017." http://web.archive.org/web/20180603182409/**http://2017study.culturetrack.com/**

CONVERSATION 12

Daniel Glaser + Takashi Kudo

"If all you see is projections of yourself, then you will never know what it is to be human."

– *Daniel Glaser*

FIGURE 12.1

Daniel Glaser (U.K.) is a self-described "brain person" with a diverse experience working at the interaction of science and arts, including as the founding director of *Science Gallery London* at King's College. Well-known as a neuroscientist for his column in *The Guardian* newspaper, Daniel was the world's first scientist-in-residence at an arts institution, at the *Institute of Contemporary Arts* in London, U.K., in 2002. Before joining the *Science Gallery London* he held multiple roles at the Wellcome Trust in Special Projects, Public Engagement and Science Engagement, again working at the nexus of science and art.[1]

Takashi Kudo (Japan) is an artist, designer and Communications Director with teamLab in Tokyo, Japan. teamLab are an artist collective who create large-scale immersive installations and exhibitions that probe technology, nature and human experience. We wanted to bring a voice from teamLab into this conversation after multiple visits to their exhibitions and upon reading about their plans to create *teamLab Borderless*, a new digital art museum in Tokyo, Japan.[2]

We brought Daniel and Takashi together due to their unique, yet overlapping, expertise where science, art and technology meet. Additionally, we were aware that Daniel had experienced teamLab installations previously and would be able to bring his own observations.

This conversation was recorded on April 20, 2018. At the time, Daniel Glaser was the Director of *Science Gallery London* **in the U.K. and Takashi Kudo was an artist, designer and Communications Director at teamLab in Tokyo, Japan**.

KEIR: Takashi, could you describe teamLab?

TAKASHI: It started in 2001 with five members whose backgrounds were in techno-physics, robotics and programming. It's now around 500 members. 80% are hands-on specialists . . . different computer language programmers, mathematicians, CG [computer graphics] animators, architects, designers, software engineers, hardware and sound engineers. We are an art collective.

We belong to the *Pace Gallery*. The artist's name is teamLab. We're not interested in individual's names, because we believe that what we are creating we cannot create without a team.

Have you ever watched *Star Wars*?

DANIEL: Mmmm (affirmative)

---

[1] The Wellcome Trust is a politically and financially independent foundation that supports scientific researchers to "take on big health challenges" and to "campaign for better science."
See: https://web.archive.org/web/20181004025515/https://wellcome.ac.uk/about-us

[2] *teamLab Borderless* opened in 2018, after this conversation was recorded. See: https://borderless.teamlab.art/

TAKASHI: *Star Wars'* director's name is George Lucas. A couple of years ago, *Frozen* or *Zootopia* were huge hits. Do you remember the director's names? People don't remember. Pixar made them.

We use the iPhone. It is very divided work. You cannot see which part an engineer made and which part a designer made. In the same way, we feel that we have to create art with teams.

Inside teamLab is not two-dimensional like hierarchies. It's more like an internet, like a three-dimensional organisation. I mentioned architects and mathematicians. Which is more important or stronger? Which is better than the others? We cannot say. When we create something it's like a territory. It's software, or it's architecture, or it's hardware.

Nobody can see everything. Instead we try to create a better prototype. If we fail, we try to find the problems. If we try to solve something, it's from within our territories. We make another prototype. One day, we feel it's finalised.

It is art that we cannot explain by words. Like a joke. If you can explain the joke perfectly, the joke is already dead.

KEIR: In this self-organising system of people with different but connected skills, how do you decide which projects to do? How do know what stories you want to tell?

TAKASHI: Because it's conceptual work we have to bring something from a higher dimension to a lower dimension. We want to explain something we cannot explain, but we feel. Most of our artworks are installations, so we try to create and build in three dimensions.

We get a schedule or deadline from the museum, and there is a theme, and if we feel it, we create. This is hard to explain, but if we can smell, we can create. We try to find a way. It's an architect thinking in an architectural way. The mathematicians, they're going to try to solve this with an algorithm. In the end, it's the process. We don't believe that we are geniuses. We have to create, and we have to create with teams. In the process of the development, we need to find some kind of core intelligence. This kind of intelligence or knowledge has become the heritage of teamLab. It's why we create everything ourselves.

If I compare to fine art. Behind one artist drawing on a canvas, there's many innovations. Hundreds of years ago paint dried quickly, so somebody invented the tube. And then the palette, easel, canvas and brush.

It's the same kind of same thing. We want to create the brush and easel ourselves, and a canvas that can be everywhere. Paint for us is light. All those things did not exist in this world, so we created them from scratch.

KEIR: With teamLab you have shared authorship. There is also shared creation at *Science Gallery*, although in a different way. Daniel, can you take us through the approach used at *Science Gallery*?

DANIEL: I'd be happy to do that, although I don't think that Takashi has answered the question of how teamLab chooses which projects to do.

TAKASHI: Oh, we don't choose.

DANIEL: You can't do things in a hundred locations around the globe at the same time, so there has to be some kind of capacity or practical limit on the number of new works you produce in a given year. I wonder whether that simply emerges, or whether there is some sort of decision structure?

TAKASHI: Our strategy right now is "yes" to all. That is our policy.

We always say "yes," but there are also human resource and budget constraints. We cannot, for example, collaborate with brands to create original artworks, because we make things that we cannot explain by words. Sometimes a client who wants to work with us on commercial advertisement has a message. If we put this message inside that creation, the output cannot be art anymore.

DANIEL: There is quite a lot of similarity between the way that you describe yourself functioning in teamLab and the brain.

When we talk about our functioning as biological entities, we describe ourselves as making "decisions." We walk through the world, we see things, we say things, we choose things, we believe things. There is this kind of mythical unity of purpose and of intention of perception. But, as we know, what we are conscious of is a very small fraction of what is happening in our brain. Most of the things that we think we are choosing deliberately are arising by some unknown and invisible internal process, which takes place outside of our awareness.

The reason that consciousness is a good idea is because you can only walk in one direction at once, right? Your body can only be in one place at once, you can only look at one thing more or less at any given second, and you can only say one word at a time. The analogy for teamLab is that you can say "yes" to as many things as you like, but you, Takashi, can only physically be at one place, and there is a finite material limit on how many works you can produce and maintain and support.

What's interesting about the way you describe teamLab functioning is that you refuse to answer the questions of how you choose what you do. . . . You describe this inevitable choice of what you *do* do (and what you don't do) as simply arising from the collective practice. You deny that choice is involved. You say you say "yes" to everything.

It's evident that you are not producing a million works a year, right? So, there is a limit. But you describe the limit as being outside the processing, as emerging from the thing. I think your account of that is closer to how the brain works than my account of *Science Gallery* is going to be, but there is a degree of honesty about it. Or, at least, the description I'm about to give you of *Science Gallery* is as dishonest as your description, because in both cases there is a mechanism which results in the choice. But I agree with you, Takashi, to describe it, to write it down as a flow chart or an algorithm, to limit it in a few words would be inaccurate.

I think that the process of making work organisationally, the process of deciding which work is made or arriving at a limited corpus of work, is an emergent property. To pin it down would be impossible, although it's very tempting to do so. As conscious beings, we try to pin our own brains down in terms of a conscious narrative.

If you forgive that digression, I'm ready to answer your question. Is it okay to carry on?

KEIR: Yes, please do.

DANIEL: There has been a *Science Gallery* in Dublin for 10 years, and that has been very helpful to me. It gives me some confidence that this thing can be achieved. It doesn't mean that I won't fail. But it means that failure is not inevitable. The second thing that we did was to over-systematise the Dublin approach to use as a weapon in cultural and corporate battles that we needed to fight within the university.

Universities don't do things like galleries very well. There are exceptions, but universities are generally too chaotic, too individualistic, too amateurish in their organisational structures to do the production of culture really well. Increasingly in the U.K., universities are becoming more corporately focused, with students as consumers and tight financial control. For both these reasons, it was helpful in starting up *Science Gallery London* to pretend that there is a fixed curatorial model that we are simply implementing.

We run seasons of activities. Everything public-facing that happens in the gallery at any given moment is part of a season. We're running one season per year in the pre-opening phase, and three per year once we're open. We choose the topics of the seasons about two years before they start, and front load a lot of the engagement work in the development period.

As soon as we have chosen a theme, we start to do engagement with researchers, artists and young people, particularly young people from non-dominant groups. They help us shape the curatorial agenda for the season. The opening season for the building is going to be called "Hooked." It's about addiction. I thought it was going to be about. . . I don't know, heroine, alcohol, this kind of thing, but when we started to talk to young people a year and a half before the season, it became clear that they were more interested in technology and social media and apps. Although there will be lots of drugs in the show, the curatorial direction is set around technology.

Once the curatorial agenda is set, we invest a lot of staff and curator time in brokering relationships between young people, artists and scientists. For example, the team have been working with children at a Secure Training Facility, which is basically a youth prison with a bunch of very marginalised young people, many of them in their early teens, who are often seen as "bad guys." They're effectively in prison at age 14 because they've been involved in drug dealing.

We've taken a couple of spoken word artists and a visual artist into that space for three or four sessions, which is going to result in some visual and spoken word work that will be in the gallery. The kids won't be able to come into the gallery because they're in prison, so this is getting voices from young people who are completely marginalised, working with a leading artist, who approached us with a proposal. We're also inviting King's College Addiction Scientists to be part of the conversation.

We've no idea what the artistic outcome of it is, but we know that it will be driven by the interests of the young people. We know it will be high artistic quality. And it's come as a result of an open curatorial process. The rigorous application of this formula is at the heart of almost everything we do.

To choose the topics, we have what we call a "Leonardo Group." When we chose the people, who would be in it, me and Jen Wong, the head of programming, we made a list, and of course we ended up with a list of old white men. So, we put these old white men in a paddock, in a virtual enclosure, and kept working. Every time we got an acceptance from a non-old, non-white, non-man that agreed to be in the group, we would release one of the old white men from the corral and invite them.

We've ended up with a group which is more than half women, about a third black and all ages. We're explicit about our commitment to young people's participation, to representing London as it is now and to a particular curatorial model. I've always been a believer in applying explicit curatorial or programmatical directorial or management approaches, because when these things are clear, I can give freedom to my curators and to do exactly what they want within those constraints.

I find the idea that you can do absolutely anything really creatively stifling.

KEIR: With both teamLab and the *Science Gallery*, there's intentionality about bringing people together, but a sort of openness to what they will create. When you're thinking about the content of your exhibition or the artwork to be created, the spaces that you generate, how important is it that the content is relevant to the people who are going to experience it?

DANIEL: In *Science Gallery*, we speak to our audience from the moment that we've chosen the topic. The only thing that we don't go out for is to choose the themes, because I subscribe to the Steve Jobs thing, "I don't ask my consumers what they want, because by the time I've made it for them, they'll want something else." But once we've chosen the theme, from the beginning, we go and engage with marginalised young people who wouldn't normally come to a gallery, along with artists and researchers. The work that we show in the exhibition arises from a deep, carefully facilitated, honest, outward-facing engagement with our audiences. That is the crucible within which the work is made by professionals.

KEIR: Fundamental is the relationship with the audience then.

DANIEL: Correct, correct. The participants.

KEIR: Takashi, when you think about what people take from a visit to a teamLab installation, is it important that they are transformed? Are you looking to transform how people see the world? See themselves? See nature? Or is about play? About fun?

TAKASHI: At teamLab, we try to make things that explore how we see the world. Things that, if people recognise and follow, it's going to change the world. We ask questions, and there is an enormous number of questions.

The thing that's most important for us is how to recognise other worlds and the relationship between humans and the world. We are living in a material world, so the material is limited. But the internet or digital is unlimited. Our artworks do not exist, they are software, and people cannot see how it works. It's light from projectors or monitors. People can be inside, but they're in our world. We believe that experience is very important, because experience is at a higher dimension than words. Only experience can change people's way of thinking or values. If we can change the values of the world, everything is going to be changed.

Digital is very flexible, it's always changing, so we can bring the people inside the artworks. Before there were boundaries. When we tried to explain something, humans had only materials. When we use materials there are going to be boundaries. There is something eternal – the boundary between the world and humans. Creatively, there is no boundary inside our minds.

As an example, when you watch a movie or television on a monitor there is a boundary between the expression and yourself. What we want to do is extend the physical world, so you're inside our artwork with your physical body, and you're a part of the artwork. Interactivity is not our core, but it is natural with digital. Our visitors are one part of the artwork. And it's playable, or fun, or beautiful. In this borderless world, when people are inside that world, and they feel positive, maybe they're going to shift. But how the visitor is going to react, we cannot control.

DANIEL: You introduced the word, "play," Takashi, when you spoke about what visitors are doing in your pieces. The achievement of building beautiful works, which people can interact with by pressing walls or with their movements, is interesting. But you seem to be downplaying the significance of the fact that the visitors are being quantified, digitised, measured and incorporated into the artwork. As if that were a mere device – "play."

How technically important to the work is the fact that the audience are part of it? Do you do user testing at an early stage in the development? How non-experts interact with your piece is critical, right? If only experts know how to make it work, it's going to fail. Bringing non-teamLab members into the development process through user testing would be a way in which the audience is playing a role. How does the user testing play into the answer to the question of how audiences help you make your work?

TAKASHI: Do we do user testings? Yes. Sometimes we use my son, or other people.

We want to use digital to expand art. For example, one of my favourite artworks is the *Mona Lisa*. If you see the *Mona Lisa* at the *Louvre*, this famous perfection of beauty, you cannot touch it because we protect this beauty.

But in front of this *Mona Lisa* there are many people, right? Sometimes this annoys us, because the relationship between human and art is one on one.

In our exhibition at *Pace [Gallery] London*, you can enter by yourself, but if there is another person, they are going to make even more flowers, and make a more beautiful landscape. After that, even if we didn't know each other, we can be a little bit positive to each other, because they are also one part of the beautiful.

DANIEL: So, if I'm watching a *Mona Lisa* or –
TAKASHI: I'm not against the *Mona Lisa* . . .
DANIEL: No, and I like it very much, but I hate people when I'm watching it. If it's me alone looking at it, then there is one experience. If there's a small number of like-minded people watching it, that's good. When there's more than a certain number, then I want to kill everybody else.

I'm re-reading John Berger's *Way of Seeing*, about the role of unique artworks in the age of digital reproduction. The divisions we make between the real world and the digital world are a bit artificial, because what we do when we're perceiving is to project meaning onto the world. With digital, what you're doing with your works is making the way in which the observer makes the work more manifest, right? If I can systematise the effect of tapping the wall, or of sitting down, as you do in teamLab, it reveals the fact that it is the individual and the other humans and the system of knowledge around us, which generates the percept. Whereas, when it's an oil painting behind glass in a gallery, you can fool yourself into thinking that there is some physical object which is hitting your retina and that what you're doing is perceiving the artwork.

In *Science Gallery*, we're making a claim about the construction of knowledge and experience and perception about science, and saying that it's quite similar to that which is about art. The conventional model is that there is a professionally mediated, elite group who control meaning in science or in art. So, scientists and artists are coterie groups. They're resistant to the idea that the public should have a say in interpreting what's what, so curators carefully guard the meaning of art – what's allowed to be art, and what isn't, and how you're supposed to behave in the presence of art. Scientists are intolerant of public's trying to reinterpret scientific results.

I'm not saying that there is no physics. I'm not saying there is no reality. But in *Science Gallery*, by adopting an explicit social justice model of empowerment, by which we bring in marginalised young peoples' voices to the construction of the artworks, we're seeking to undermine the control that experts exert over both artistic and scientific discourse.

We're not telling the experts that we're doing that. We're selling them the idea of engagement and empowerment and participation. But actually, I think we're trying to subvert the control.

This is not community art. We're not saying there is no expertise in artistic production or in curation. What I'm claiming is that we'll make better art, and better science, if we can open the production process up to a wider range of voices and inputs, while still recognising, celebrating and ultimately giving control to expert curators and artists.

KEIR: There is an idea about our digitally transformed world that people are both more connected, and at the same time feeling lonelier – "together, alone."

I've found a lot of the modern technologies that I engage with seem to be ever so cleverly isolating me, or understanding me so well as to cater to me almost perfectly, created by what Jaron Lanier calls behaviour modification empires. They frame around me a Keir-shaped box, a digital sarcophagus, training me to using them so that I am isolated from the people that are physically proximate.

When I've experienced teamLab, which I've done in three different contexts now, I want to put my phone away and be really present with strangers. I want to play with the work and the other people. I want to complete the work.

Daniel, as you plan out the opening of *Science Gallery*, how important is it that people are in dialogue with each other, and with the work?

DANIEL: There's a concept developed by James J. Gibson of an "affordance." An affordance is a thing that makes you want to do to it. For instance, if you have a very fine tea cup, you reach out for it with a very fine grip. If you have a big tankard of beer, you reach for it in a different way. There are systems in the brain that specifically look at objects and shape our movements and our approaches to them in a particular way.

The teamLab environments that I've experienced make people behave in particular ways. They make people quiet. They make people move slowly. They make Keir put his phone away. They make people curious about what each other are doing. They make people look where others have been, not just where they're going. In other words, they produce behaviours in people.

But it has to be important in your development process that the expert developers aren't the only people interacting with the thing you're building, because you're going to get all kinds of combinatorics from novices using the technology in the gallery. It can't be the first time that a non-expert uses it is when it's been in the gallery. By contrast, you could imagine that a painting that's seen by the public for the first time when it's put on the gallery wall, and that's probably okay. But you need non-expert interactions with your systems, Takashi, long before they are shown, because otherwise you've got no idea whether it's going to work or not.

TAKASHI: Today to make our artwork we need hardware like projectors, monitors or LEDs. This is a problem of today's technology limitations. Maybe in 20 years, we can do more, with better hardware.

Today when I go to the cinemas, I have to sit down and watch because it's displayed for that perspective. We have to sit, otherwise it doesn't work.

There is no vanishing point. But most of our artworks are based on Japanese-style drawings, which are flat.

There are so many ways to bring 3D to 2D on the canvas. Maybe we can reuse, re-create or reconstruct old Japanese-style drawings through digital technology in a three-dimensional way. After that, maybe we are inside visual art installations, or using our full bodies even for movies, not only sitting down.

DANIEL: Takashi, I'm full of admiration for the way that you bring a drawing style into an account of why your immersive environments may be more satisfying than cinema. It's beautiful. The story I want to tell is a complimentary kind of human activity.

We employ a freelance curator producer for each season. This is an important part of the model. If we're doing three seasons a year, they have to look and feel completely different. We have an in-house Head of Programming, then below the curator we have In-House Production Assistant and Gallery Manager. But we have a different freelance curator for each season, and they're engaged at intervals over the two-year process. For *Blood*, we had a curator called Andy Franzcowiak. He came to us three months before the season opened in a gallery in South London, in an artistically vibrant but also very deprived area called Peckham. He was concerned because the borough that we're in is 40% black and we didn't have many black artists involved in the work, so it wasn't representative.

He commissioned a work that ended up being called *Tough Blood*. The work is about Sickle Cell Disease, which is a genetic disease that particularly affects people of African origin. It can produce terrible crises and effects for some people who have both copies of the gene. He commissioned a filmmaker to make footage for a short documentary about the origins of Sickle Cell and about the communities where it's currently predominant in London, which is the black/African and Caribbean communities. He then commissioned a choreographer to make a piece of dance, which was performed by a black dancer, to some specially commissioned music.

The dance was performed in the gallery live, with the film projected onto the dancer. The audience was 80–90% black Afro-Caribbean, members of the local community. The evening was emceed by a young black woman from Birmingham who's got Sickle Cell. She introduced herself by saying, "My name is Jenica Leah and I'm a Sickle Cell warrior." The event was videoed and the video documentation was shown in the gallery for the rest of the season.

There is a lot of stigmatisation of those with Sickle Cell. Sickle Cell warriors need huge amounts of painkillers, but when they come to a hospital, when they're on holiday and say, "I need morphine, I need it now," hospitals often assume they're junkies and throw them out. There's a lot of ignorance, there's a lot of stigma. This artwork arose as a result of a commissioning process, which started from a narrow conception of representation of a community and of a lack of representation artists of colour, in relation to the audience we were interested in. It gave rise to a work that could not have been made otherwise. For me, it feels as if this is the *Science Gallery* equivalent of the user-embodied process which teamLab are doing by bringing the physical bodies of the audience into the work using sensors and projection, and so on.

It's the same thing. We have to find ways of bringing people into the making of the work that we're doing.

KEIR: Especially when we want to bring people together in the consumption of that work and affect how they see themselves and see the world. From a practical point of view, is there anything in our biology or our neurobiology, such that we learn differently in groups when we have these experiences?

DANIEL: Yeah, definitely.

KEIR: What are the mechanics of that?

DANIEL: Consciousness arises through interaction, but we become conscious when we see other people as agents. It's probably impossible to become conscious if you're in your "digital sarcophagus." If all you see is projections of yourself, then you will never know what it is to be human. Learning through mimicry, through interaction, through observing the movements and the understanding and the responses of others is critical.

Learning happens much faster when you can see the mistakes that other people are making.

KEIR: I remember being in one of the teamLab spaces and a young girl ran up to me and said, "You need to use your arms more." I started waving my arms and flowers started to grow on me and on my kids as they ran around. That girl seemed so engaged and powerful, because she could train this adult stranger how to actualise the work. That stuck with me.

DANIEL: I've observed that a lot in the teamLab things I've seen. Within *Science Gallery*, because we have mediators who are 15–25-year olds, many of them students, it's not like a conventional art gallery where it's a silent, contemplative space. You can expect our mediators to come up to you and say, "Hey, have you seen this really cool thing?"

They're not trained art historians, they're not experts. They're people just like the audience. What we hope is to catalyse engaged interactions, and I take this to be an empowerment process.

KEIR: What are the most replicable elements of your practice? What can staff from small and medium-sized institutions take from your practice to replicate and present to their publics?

DANIEL: I think the thing that people can take from what we do is this commitment to social justice. People talk about "hard to reach" audiences. I think that's bullshit. I don't think the audiences that you're interested in are hard to reach. In fact, one of the features of being a young person with social difficulties is that your movement is restricted. If young people involved in crime move more than a few hundred metres from their house, they're in danger of being attacked. Many of the people from poor areas in London live less than one mile from the river, but have never been to the river.

Like the recipe I gave for having an advisory group that's not all white… This is not complicated. It requires a commitment to the outcome, then you implement a simple procedure and you end up with a group that looks like the audience you're interested in.

Go out into the communities which you're interested in. You need an invitation, you need gatekeepers and representatives from that community, but that's not hard. There are galleries there, there are video makers, there are technologists, there are educational institutions. Find that link, and make work in the space that is owned and controlled by the audience you're interested in.

You can bring your artists and your curatorial values, and you can even show the work in your shiny gallery in the end, but unless you find a mechanism to have a conversation with the people that you're interested in, in a space and in a language which they feel comfortable, you'll be trapped within the established artistic canon. To make new work, you need to find ways for young people to express themselves. Again, this is not to deny the significance of the canonical experts. I'm not saying we need to throw away science and make it anew. I'm not saying physics can be done by anyone, in the bedroom. I'm saying that you need to have expertise, including classical experts, but you want to take them into spaces, sometimes, where they are not comfortable and give space for a different audience to speak.

KEIR: I have one last question. Close your eyes and think, 10 years, 20 years from now, what sort of experiences will be available to us that blend science, nature and art together? What would you love to try to do?

TAKASHI: I really don't know about tomorrow; I cannot see the future.

Right now, we are creating our own museum in Tokyo. It is 10,000 square meters – a museum that's small in size, and the whole building is one digital art piece. That is some part of our future vision. It's very hard to explain with words.

We don't know what's going to happen.

DANIEL: If you look at the history of music, until very recently, music was much more participative than it is now. A hundred years ago, most people's experience of music was by making music. I grew up with that. My grandfather and my mother played in a classical string quartet and I have to say, they were terrible. The quality of the performance was awful. Nobody would pay to listen, but they experienced Mozart's string quartets by playing them.

I find my children watching YouTube videos of people playing Minecraft. This is terrible, because it's not curated, it's not edited or produced. It's the most banal bullshit. What I hope for in 10 years' time is that people are making more work, more art. That their experience of culture is through production.

You know this idea that everyone's an artist? I think that's bullshit, too. I think there's a small proportion of us, not including me, who are artists. But being involved in the production of art is becoming more of an elite pursuit. In Britain, only the rich schools have music teachers, with the increased emphasis on technical learning and the curriculum. There's less and less space for creativity within the schools. We need to have people who are producers of culture, not just consumers. That's not saying that culture becomes user-generated and there are no more galleries and no more artists. We still need high culture; we still need expertise. But people need to be looking at drawings in galleries because they themselves are doing drawings.

TAKASHI: Yes.

DANIEL: That's my hope for 10 years from now. And galleries can be part of that.

TAKASHI: I agree with that. I didn't say everybody is an artist, I just said the people can be part of art. I don't know if we are artists even. But I very much understand your point of view.

KEIR: Thank you so much, Takashi. Thank you so much, Daniel. This has been a treat.

Throughout this book, we have argued that museums must be inclusive, polyvocal and responsive to their communities. However, collaborative work brings many challenges, including questions about shared authorship and ownership. This conversation with Daniel and Takashi offers some useful provocations on these topics. When developing collaborative projects, each participant should seek to solve the problems at hand from within their own territories, using their own specific strengths and expertise, which may include lived expertise. Relationships are, as we've often discussed in this book, essential to the success of any museum project and the project of museum making, and should be carefully facilitated and enabled. As Daniel notes, "you need an invitation, you need gatekeepers and representatives [from the communities you're working with]."

This conversation also maps alternative curatorial strategies and methodologies for creating exhibitions. For more on this topic, we suggest you read:

Conversation 4: Sarah Brin + Adriel Luis
Conversation 11: Tony Butler + Lori Fogarty

DIGITAL PRACTICE IN MUSEUMS: WHERE DO WE GO FROM HERE?

In the final decade of the 20th century, Stephen E. Weil famously argued that museums were transforming, "from being about something to being for somebody" (Weil 1999). While exploring a new emphasis on the museum's role as a public service organisation, Weil described an obligation for museums to be responsive to their communities, "to enrich the quality of individual lives and to enhance their community's well-being" (Weil 1999, 255). Today, as reflected in the conversations herein, such ideas have developed to the extent that museums are now *expected* to be responsive to their communities, measuring their impact upon and within their communities far beyond the boundaries of the institutional walls. Indeed, John Fraser, editor of *Curator: The Museum Journal*, recently argued that, "there is no longer a new museology, but rather, a museological practice that continues to advance through experimentation, testing, and eventually evaluation to assess improvement" (Fraser 2019, 96). Here, Fraser uses language similar to that of agile digital product development, a process frequently cited by participants in this publication. Concurrently and necessarily, digital practice in museums has been maturing such that it has become normative and expected. Within the post-digital museum, a term coined by Ross Parry, the dominant discourses of change and progress are subsumed to make room for deeper and more challenging questions, so that, "we can be free to reach for alternative sets of theoretical reference points, and break away from the gravitational pull of dominant theories of technological adoption" (Parry 2013, 37).

If both of these observations are true, and we have moved to a time when museological practice takes a community-focussed, socially responsive and digitally informed approach as assumed, rather than at the center of a battle for what *should be*, an argument supported by the conversations in this book, as museum practitioners we now face a more nuanced and interesting set of questions: How can we respect the specificity of all the different micro- and subcultures that exist within a specific community, whilst also connecting them to broader histories of material and digital culture, to paraphrase Lara Day?[1] How do we recognise and appropriately compensate the often invisible, hidden labour of experience-based and lived expertise when working with knowledge communities? How can we create knowledge systems, including collections management systems, that support transsystemic knowledge,[2] and enable meaningful kinds of consent and agency regarding how such knowledge and objects are used?[3] How do we create anti-oppressive and welcoming online spaces for, and reflective of, all the members of our communities? How do we challenge, rather than further embed, institutional biases in our collections, social media practices and other digital spaces? If museums are now post-digital, how do we respond to the "digital dilemma" as expressed by Arthur Cohen?[4]

There are many examples of these questions being asked and this work being done, both within this book and beyond it. For instance, Jen Ross recently analysed digital co-production through a lens of modalities theory, arguing that digital co-production, "unfolds across multiple times and spaces; involves the 'unknowable other'; challenges the stability of relationships; [and] invites a rethinking of hospitality" (Ross 2018, 576). Relatedly, in Conversation 2, Barbara Makuati-Afitu speaks of the institutional challenges of being hospitable due, in part, to the complexities of power relationships, wherein the museum is the host to an outside community that has or maintains the knowledge that the institution wishes to access, systematise and hold onto within its institutional walls.

One way to address this challenge is for institutions to be truly of, by and for all their communities. OF/BY/FOR ALL, an initiative started by Nina Simon based on work conducted at the *Santa Cruz Museum of Art and History* (the MAH) which Shelley Bernstein joined as Head of Product/CTO a few months after recording her conversation for this book, seeks to help civic and cultural organisations become of, by and for their communities in order to strengthen communities and empower individuals ("Vision — OF/BY/FOR ALL" 2018). Driving their work is a belief that:

---

1 See Conversation 3: Lara Day + David Smith
2 This is a concept and term borrowed from the Faculty of Law at McGill University, whose location in "a bilingual city in a bilingual and bijural province" ("Transsystemic Legal Education | Paul-André Crépeau Centre for Private and Comparative Law - McGill University" n.d.) in Montreal, Québec, prompted a new, dialogic approach to the teaching of law that reflects the coexistence of multiple legal cultures. It emphasises relations, not merely between legal jurisdictions, but also around cultures, philosophies, interpretive approaches and even individuals (Praagh 2005). In this case, the term is used to consider the multiple different and plural systems of knowledge, including local and Indigenous knowledges, that museums are expected to deal with in the 21st century.
3 Traditional Knowledge (TK) Labels, which are digital labels designed to enable indigenous communities to add existing local protocols for access and use to recorded cultural heritage circulating outside community contexts, are a useful tool for this purpose. See https://web.archive.org/web/20190714164925/**http://localcontexts.org/tk-labels/**
4 See Conversation 10: Arthur Cohen + Tonya Nelson

if you want to be FOR your whole community, you have to be representative OF them and co-created BY them. If people don't see themselves as part of your work, they won't see your work as an essential part of their lives.

("Vision — OF/BY/FOR ALL" 2018)

Similarly, the "Creating Anti-Oppressive Spaces in Museums" project includes tools that museum workers can use to make their institutions as inclusive as possible. The project, run by Sina Bahram, Eric Gardner, Sarita Hernandez, fari nzinga, Trish Oxford, nikhil trivedi and Aletheia Wittman, includes a series of checklists, ideas, and definitions to help equip practitioners and institutions make anti-oppressive projects.[5]

Emerging are examples of museums actively addressing their previous reliance of dominant cultural narratives and seeing immediate success in attracting audiences who previously did not feel welcome enough to attend. The *High Museum* in Atlanta, U.S. tripled its non-white audience in the two years to 2017 through a whole museum approach (Halperin 2017). The *Manchester Museum*'s 2021 expansion project has been branded *hello future,* as they seek to become "more inclusive, imaginative and relevant to the diverse communities we serve" by including a South Asia Gallery, a Chinese Culture Gallery and the world's first Centre for Age Friendly Culture in their small physical transformation (Ward, 2018). Foundations funding museums in the U.S. are also taking note. In 2017, the Ford and Walton Family Foundations launched the Diversifying Art Museum Leadership Initiative, followed by a similar initiative from the American Alliance of Museums a year later, and in 2018, the Knight Foundation announced funding for eight digital-focused positions at museums in the U.S., as well as funding for museum technology prototypes to help institutions embrace technology and overcome the resource roadblocks they currently face.

And, as museum professionals with a strong background in technology begin to move from manager and executive to director level positions, we believe change in the sector will accelerate. In early 2019, the *Hunt Museum* in Limerick released its ambitious 2025 strategy in which they seek to make a positive social and economic impact through their operations in the physical, virtual and community domains. Jill Cousins, previously a co-founder of the massively successful digital culture initiative Europeana and now Director and Chief Executive of the *Hunt Museum*, writes that the museum "will fully integrate the virtual and physical worlds of the museum to cater to both online and offline visitors" as they seek to "make cultural heritage a keystone in Digital Strategy regionally and nationally, jump started by the launch of Limerick as Ireland's first digital city" (Hunt Museum 2019).

---

5 The "Creating Anti-Oppressive Spaces in Museums" project can be found at https://web.archive.org/web/20190714164651/**https://github.com/the-incluseum/anti-oppressive-spaces**

To be a museum digital practitioner today means engaging with complex and often difficult social, political, economic and technological issues in order to genuinely understand and respond to the multiple environments within which we're working. As individuals and actors within the museum industrial complex, we must be thoughtful about our practice and aware of the implications of the choices that we make in considering partnerships with technology companies, collecting and storing audience data, uncritically digitising collections and making them available online without also considering the power relations at stake in doing so. There are also risks in assuming that online spaces are accessible, democratic and equitable by default. Queering and Indigenising museum staff, content and collections needs to be supported by the technologies that underpin and syndicate that work. As discussed in the opening chapter, institutional practices are replete with contradictions, and often risk further embedding institutional biases and systemic oppression even as they seek to challenge them. Languages, cultures and abilities that are not deliberately included and supported in online spaces risk falling into digital decline, (Kornai 2013) and the decisions we make about which stories, histories, languages and objects to prioritise have flow on effects.

These kinds of questions ask museum digital practitioners to consider more fully whether the work they do enhances the well-being of their communities, as well as supporting the mission. Doing so changes calculations of risk and reward, shifting the focus from concerns about institutional risk to consideration of societal risk, and asks us to imagine who is benefitting from our work and who might be harmed by it. As Adriel Luis comments in Conversation 4:

> It's basic practice to make sure that the museum isn't constructed to put its community in harm's way. But when we build websites or online exhibitions, or we run our social media, we don't consider that digital space with the same vigilance. When Snapchat first came out, I don't know how many museums read the terms and services before deciding to join and try to get their visitors onto it, too. It's not about Snapchat specifically, it's not about Facebook, it's not about Uber, it's not about any individual company and whether or not I trust that company. It's about where museums are directing people to in digital space, and whether museums understand the environment within which they're leading their communities.

It is also true that we cannot ignore the logic of accumulation and interconnection that defines our time, and the intimate linking of digital and surveillance technologies with capitalism. Koven Smith argues,

> It has become clear that the stated values of [commercial software] companies (connecting people, sharing ideas) often have little to do with the actual values embodied in their products (monetising user data, invading privacy, enabling harassment, and so on). This means that the values and

principles inherent in the technology itself are diverging sharply from the values of the museums using it. It is therefore time for a reckoning: we must now address not just the *practical* considerations of the technology we use, but also its *moral and ethical* implications. If we don't, we risk compromising the values of the museums we serve.

(Smith 2019)

The digital future of museums, therefore, asks both institutions and practitioners to be critically engaged, ethically informed and values-led. Seb Chan has argued, "we *need* museums to be democratic spaces, to function as engines of curiosity for our communities, and to provide important persistence in this age of ephemerality" (Chan 2019). We fully agree. We cannot escape power relations, nor the influence of the connected and interlinked histories and structures that have informed our institutional practice to date. However, by examining and acknowledging the complex, contradictory context in which our work is situated, and adapting accordingly, we believe that we can create better institutions that enhance the well-being of all members of their communities.

Courtney Johnston, who wrote the foreword to this book, once proposed that:

> those of us who have come of age as museum tech professionals have a unique set of values and approaches to our work that can form the bedrock of the modern museum: *Usability, Accessibility, Transparency, Openness, Experimentation, Craft, Sharing, and Creating value together.*
>
> *(Johnston 2015)*

These values and approaches are on display throughout the conversations in this book, and are certainly core to much of the thinking and discourse that has driven digital practice within museums over the last half century, which has made space for innovative and important work. These values continue to be relevant. However, for museums to be meaningful to their communities and publics, and resilient to a changing, uncertain world, we would add that digital museum practice should seek to bring nuance and complexity where museums have previously promoted clear categories, regularity, simplicity and certainty. Digital museum futures should be experiential and experimental; mission-driven and community co-designed; responsive, inclusive and open; evaluated and accountable; detailed and polyvocal; grounded in kindness, empathy and humility; staffed by practitioners with diverse backgrounds and experiences; and, most of all – relevant and impactful.

References

Bahram, Sina, Eric Gardner, Sarita Hernandez, fari nzinga, Trish Oxford, Nikhil Trivedi, and Aletheia Wittman. 2016. "Anti-Oppressive-Spaces: Resources for the 2016 Museum Computer Network Panel 'Creating Anti-Oppressive Spaces Online'." The Incluseum. 2016. https://github.com/the-incluseum/anti-oppressive-spaces.

Fraser, John. 2019. "Useful Museums." *Curator: The Museum Journal* 62 (2): 95–6. https://doi.org/10.1111/cura.12313.
"Hello Future | Manchester Museum." n.d. Accessed September 23, 2019. https://www.museum.manchester.ac.uk/about/hellofuture/.
Hunt Museum. 2019. "Hunt Museum Strategy 2025: Changing Lives with Culture, Creativity and Learning." https://web.archive.org/web/20190714165508/**http://www.huntmuseum.com/about-us/**.
Johnston, Courtney. 2015. "Best of 3: October 2015." Best of 3. 2015. https://web.archive.org/web/20190714170044/**http://best-of-3.blogspot.com/2015/10/**.
Kornai, András. 2013. "Digital Language Death." Edited by Eduardo G. Altmann. *PLoS ONE* 8 (10): e77056. https://web.archive.org/web/20190714165623/**https://journals.plos.org/plosone/article?id=10.1371/journal.pone.0077056**.
Parry, Ross. 2013. "The End of the Beginning." *Museum Worlds* 1 (1): 24–39. https://doi.org/10.3167/armw.2013.010103.
Praagh, Shauna Van. 2005. "Preface Navigating the Transsystemic: A Course Syllabus." https://www.mcgill.ca/centre-crepeau/files/centre-crepeau/vanPraagh_Navigating.pdf.
Ross, Jen. 2018. "Casting a Line: Digital Co-Production, Hospitality and Mobilities in Cultural Heritage Settings." *Curator: The Museum Journal* 61 (4): 575–92. https://doi.org/10.1111/cura.12280.
Smith, Koven. 2019. "At What Cost?" Koven J. Smith. 2019. https://web.archive.org/web/20190714165344/**https://kovenjsmith.com/post/the-values-of-museum-technology/**.
"Transsystemic Legal Education | Paul-André Crépeau Centre for Private and Comparative Law – McGill University." n.d. McGill.CA. Accessed March 27, 2019. https://web.archive.org/web/20190714165146/**https://www.mcgill.ca/centre-crepeau/projects/transsystemic**.
"Vision — OF/BY/FOR ALL." 2018. OF/BY/FOR ALL. 2018. https://www.ofbyforall.org/vision.
Weil, Stephen E. 1999. "From Being about Something to Being for Somebody: The Ongoing Transformation of the American Museum on JSTOR." *Daedalus* 128 (3): 229–58.

BIBLIOGRAPHY

"2017 Edelman Trust Barometer-Executive Summary." https://www.edelman.de/fileadmin/user_upload/Studien/2017_Edelman_Trust_Barometer_Executive_Summary.pdf.

Aked, Jody, and Sam Thompson. 2011. "Five Ways to Well-Being: New Applications, New Ways of Thinking." New Economics Foundation. http://web.archive.org/web/20190620111351/**https://neweconomics.org/2011/07/five-ways-well-new-applications-new-ways-thinking**

Ali, Muhammad, Piotr Sapiezynski, Miranda Bogen, Aleksandra Korolova, Alan Mislove, and Aaron Rieke. 2019. "Discrimination through Optimization: How Facebook's Ad Delivery Can Lead to Skewed Outcomes," April. https://web.archive.org/web/20190712191030/**https://arxiv.org/abs/1904.02095**.

Anderson, Susan. 2019. "Visitor and Audience Research in Museums." In *The Routledge Handbook of Museums, Media and Communication*, edited by Kirsten Drotner, Vince Dziekan, Ross Parry, and Kim Christian Schrøder, 1st ed., 80–95. Abingdon, OX: Routledge.

Armstrong, Harry, Hasan Bakhshi, John Davies, Georgia Ward Dyer, Paul Gerhardt, Celia Hannon, Svetlana Karadimova, Sam Mitchell, and Francesca Sanderson. 2018. "Experimental Culture A Horizon Scan Commissioned by Arts Council England." www.artscouncil.org.uk.

Asia Art Archive. 2017. "About the Asia Art Archive." aaa.org.hk. https://web.archive.org/web/20171219034636/**https://aaa.org.hk/en/about/about-asia-art-archive**

Autry, La Tanya. 2017. "Changing the Things I Cannot Accept: Museums Are Not Neutral." Artstuffmatters. https://web.archive.org/web/20190712191207/**https://artstuffmatters.wordpress.com/2017/10/15/changing-the-things-i-cannot-accept-museums-are-not-neutral/**.

Barlow, John Perry. 1996. "A Declaration of the Independence of Cyberspace." Electronic Frontier Foundation. https://web.archive.org/web/20190712191309/**https://www.eff.org/cyberspace-independence**.

Battaglia, Andy. 2018. "The ARTnews Accord: Aruna D'Souza and Laura Raicovich in Conversation." Art News. https://web.archive.org/web/20190712191409/**http://**

www.artnews.com/2018/05/14/artnews-accord-aruna-dsouza-laura-raicovich-conversation/.

Baym, Nancy K., and danah boyd. 2012. "Socially Mediated Publicness: An Introduction." *Journal of Broadcasting & Electronic Media*. Abingdon, OX: Taylor and Francis Group, LLC. http://www.tandfonline.com/doi/full/10.1080/08838151.2012.705200

Bennett, Tony. 2004. "The Exhibitionary Complex." In *Grasping the World: The Idea of the Museum*, edited by Donald Preziosi and Claire Farago, 413–42. Hants, UK & Burlington, VT: Ashgate.

Bernstein, Shelley. 2018. "Using Authority of the Resource Technique at the Barnes." Medium.com. http://web.archive.org/web/20180912172131/**https://medium.com/barnes-foundation/using-authority-of-the-resource-technique-at-the-barnes-e6dca41f1b62**

Black, Graham. 2016. "Remember the 70%: Sustaining 'Core' Museum Audiences." *Museum Management and Curatorship* 31 (4): 386–401. https://doi.org/10.1080/09647775.2016.1165625.

Blanco, Lina. 2018. "Google Arts and Culture #Selfie App Inherits Art World Disparities." KQED Arts. January. http://web.archive.org/web/20190617123335/**https://www.kqed.org/arts/13819312/google-arts-and-culture-selfie-app-inherits-art-world-disparities**.

Bolander, Elizabeth, Hannah Ridenour, and Claire Quimby. 2018. "Art Museum Digital Impact Evaluation Toolkit." Office of Research & Evaluation, Cleveland Museum of Art.

boyd, danah. 2017. "Did Media Literacy Backfire?" Data & Society: Points. https://web.archive.org/web/20190712191604/**https://points.datasociety.net/did-media-literacy-backfire-7418c084d88d?gi=3f5228e95dae**.

———. 2018. "You Think You Want Media Literacy… Do You?" Data & Society: Points. https://web.archive.org/web/20190712191646/**https://points.datasociety.net/you-think-you-want-media-literacy-do-you-7cad6af18ec2?gi=8b52512e5eaa**.

boyd, danah, and Kate Crawford. 2011. "Six Provocations for Big Data." *Oxford Internet Institute's "A Decade in Internet Time: Symposium on the Dynamics of the Internet and Society."* Oxford, UK. https://web.archive.org/web/20190713073737/**https://papers.ssrn.com/sol3/papers.cfm?abstract_id=1926431**.

Buolamwini, Joy, and Timnit Gebru. 2018. "Gender Shades: Intersectional Accuracy Disparities in Commercial Gender Classification." In *Proceedings of Machine Learning Research*, edited by Sorelle A. Friedler and Christo Wilson, 81:1–15. Conference on Fairness, Accountability, and Transparency. http://proceedings.mlr.press/v81/buolamwini18a/buolamwini18a.pdf.

Cairns, Susan. 2013. "Mutualizing Museum Knowledge: Folksonomies and the Changing Shape of Expertise." *Curator: The Museum Journal* 56 (1): 107–19.

Cameron, Fiona, and Sarah Mengler. 2009. "Complexity, Transdisciplinarity and Museum Collections Documentation: Emergent Metaphors for a Complex World." *Journal of Material Culture* 14: 189–218.

———. 2010. "Activating the Networked Object for a Complex World." In *Handbook of Research on Technologies and Cultural Heritage: Applications and Environments*, edited by Georgios Styliaras, Dimitrios Koukopoulos, and Fotis Lazarinis, 166–87. Hershey, PA: IGI Global. https://doi.org/10.4018/978-1-60960-044-0.

Carvin, Andy. 2005. "Tim Berners-Lee: Weaving a Semantic Web." A Sense of Place Network. http://web.archive.org/web/20100606132725/**http://www.cbpp.uaa.alaska.edu/afef/weaving%20the%20web-tim_bernerslee.htm**

Chan, Seb. 2019. "Fire, Fire, Fire—Words from Creative State 2019." Noteworthy – The Journal Blog. https://web.archive.org/web/20190712191910/**https://blog.usejournal.com/fire-fire-fire-words-from-creative-state-2019-b314f33da1c4?gi=f51e8fd97758**.

Christen, Kimberly, Alex Merrill, and Michael Wynne. 2017. "A Community of Relations: Mukurtu Hubs and Spokes." *D-Lib Magazine* 23 (5/6). https://doi.org/10.1045/may2017-christen.

Cope, Aaron Straup. 2018. "Successful Distractions." [This Is Aaronland]. https://web.archive.org/web/20190712191945/**https://www.aaronland.info/weblog/2018/09/12/distractions/**.

Cormier, Brendan. 2017. "How We Collected WeChat." V&A Blog. September. http://web.archive.org/web/20190620070332/**https://www.vam.ac.uk/blog/international-initiatives/how-we-collected-wechat**.

Daley, Paul. 2017. "Songlines at the NMA: A Breathtaking Triumph of 21st Century Museology." *The Guardian*, September 15, 2017.

"Democracy Index 2018: Me Too? Political Participation, Protest and Democracy." 2019. London. https://web.archive.org/web/20190630082129/**https://www.eiu.com/public/topical_report.aspx?campaignid=Democracy2018**.

Department for Digital, Culture, Media & Sport, U.K. Government. 2018. "Culture Is Digital." March 2018. https://www.gov.uk.

Derby Museums. 2018. "Derby Museums Business Plan." Derbymuseums.org. https://www.derbymuseums.org/wp-content/uploads/2018/03/0383_DM-Business-Plan_A5-12pp_digital.pdf.

"Digital Culture 2017." London. https://www.artscouncil.org.uk/sites/default/files/download-file/Digital Culture 2017_0.pdf.

Dilenschneider, Colleen. 2017. "People Trust Museums More than Newspapers. Here Is Why that Matters Right Now." Know Your Own Bone. April 2017. http://web.archive.org/web/20170815024834/**https://www.colleendilen.com/2017/04/26/people-trust-museums-more-than-newspapers-here-is-why-that-matters-right-now-data/**.

Dimock, Michael. 2019. "An Update on Our Research into Trust, Facts and Democracy." Pew Research Center. https://web.archive.org/web/20190712192126/**https://www.pewresearch.org/2019/06/05/an-update-on-our-research-into-trust-facts-and-democracy/**.

Doctorow, Cory. 2017. "Three Kinds of Propaganda, and What to Do about Them / Boing Boing." BoingBoing. https://web.archive.org/web/20190712192224/**https://boingboing.net/2017/02/25/counternarratives-not-fact-che.html**.

Dunn, Brad. 2017. "Look at the Entire Customer Journey 'From Couch to Couch'." CMS Wire. October 2017. http://web.archive.org/web/20171208130139/**https://www.cmswire.com/digital-experience/brad-dunn-look-at-the-entire-customer-journey-from-couch-to-couch/**

Dunne, Carey. 2016. "Danish National Gallery Removes the Word 'Negro' from 13 Artworks' Titles." Hyperallergic. https://web.archive.org/web/20190712192317/**https://hyperallergic.com/304385/danish-national-gallery-removes-the-word-negro-from-13-artworks-titles/**.

Frangonikolopoulos, Christos A., and Ioannis Chapsos. 2012. "Explaining the Role and the Impact of the Social Media in the Arab Spring." *Global Media Journal: Mediterranean Edition*, Fall: 10–20. https://web.archive.org/web/20171130171857/**http://www.academia.edu/2370755/Explaining_the_role_and_impact_of_social_media_in_the_Arab_Spring_**.

Fraser, John. 2019. "Useful Museums." *Curator: The Museum Journal* 62 (2): 95–6. https://doi.org/10.1111/cura.12313.

Funtowicz, Silvio, and Jerry Ravetz. 2001. "Post-Normal Science. Science and Governance under Conditions of Complexity." In *Interdisciplinarity in Technology Assessment*, 15–24. Berlin, Heidelberg: Springer Berlin Heidelberg. https://web.archive.org/web/20180605115518/**https://link.springer.com/chapter/10.1007%2F978-3-662-04371-4_2**.

Gil, Britney. 2016. "Texts, Facts, Emotions, and (Un)Making a Nation – Cyborgology." Cyborgology. https://web.archive.org/web/20190712192418/**https://thesocietypages.org/cyborgology/2016/12/02/texts-facts-emotions-and-unmaking-a-nation/**.

Goggin, Benjamin. 2018. "Is Google's Arts And Culture App Racist? – Digg." Digg. http://web.archive.org/web/20190712192454/**http://digg.com/2018/google-arts-culture-racist-face**.

Grinell, Klas. 2014. "Challenging Normality: Museums In/As Public Space." In *Museums and Truth*, edited by Annette B. Fromm, Per B. Rekdal, and Viv. Golding, 169–88. Newcastle upon Tyne, England: Cambridge Scholars Publishing.

Gurian, Elaine Heumann. 2017. "Complexity Theory and Museums 2017, a Speech for MuseumNext Melbourne 2017." http://www.egurian.com/omnium-gatherum/issues/civility/complexity-theory-and-museums-2017.

Halperin, Julia. 2017. "How the High Museum in Atlanta Tripled Its Nonwhite Audience in Two Years". Artnet news. http://web.archive.org/web/20180417044636/https://news.artnet.com/art-world/high-museum-atlanta-tripled-nonwhite-audience-two-years-1187954.

HUD Public Affairs. 2019. "HUD Charges Facebook with Housing Discrimination Over Company's Targeting Advertising Practices." Washington, DC: US Department of Housing and Urban Development. https://web.archive.org/web/20190712192623/**https://www.hud.gov/press/press_releases_media_advisories/HUD_No_19_035**.

Hunt Museum. 2019. "Hunt Museum Strategy 2025: Changing Lives with Culture, Creativity and Learning." https://web.archive.org/web/20190714165508/**http://www.huntmuseum.com/about-us/**.

Johnston, Courtney. 2016. "Final Report: Visitor Experience in American Art Museums." Best of 3. http://web.archive.org/web/20181001033400/**http://best-of-3.blogspot.com/2016/06/visitor-experience-american-art-museums23.html**

———. 2015. "Best of 3: October 2015." Best of 3. https://web.archive.org/web/20190714170044/**http://best-of-3.blogspot.com/2015/10/**.

Kane, Natalie D. 2019. "No Title." Twitter. https://web.archive.org/web/20190712192728/**https://twitter.com/nd_kane/status/1144195380076302336**.

Kenderdine, Sarah. 2013. "'Pure Land': Inhabiting the Mogao Caves at Dunhuang." *Curator: The Museum Journal* 56 (2): 199–218. https://onlinelibrary.wiley.com/doi/abs/10.1111/cura.12020.

Kornai, András. 2013. "Digital Language Death." Edited by Eduardo G. Altmann. *PLoS ONE* 8 (10): e77056. https://web.archive.org/web/20190714165623/**https://journals.plos.org/plosone/article?id=10.1371/journal.pone.0077056**.

Kuhn, Thomas S. 1962. "The Structure of Scientific Revolutions." In *Foundations of the Unity of Science*, 2nd ed., edited by Otto Neurath, 59. Chicago and London: The University of Chicago Press.

Lackoi, Krisztina, Maria Patsou, and Helen Chatterjee et al. 2016. "Museums for Health and Wellbeing. A Preliminary Report, National Alliance for Museums, Health and

Wellbeing." https://museumsandwellbeingalliance.files.wordpress.com/2015/07/museums-for-health-and-wellbeing.pdf.

LaPlaca Cohen. 2017. "Culture Track 2017." http://web.archive.org/web/20180603182409/**http://2017study.culturetrack.com/**

"Law Professor Frank Pasquale Q&A · Nesta." 2019. Finding Ctrl. https://web.archive.org/web20190712193057/**http://findingctrl.nesta.org.uk/frank-pasquale/**.

Lewis, Rebecca. 2018. "Alternative Influence: Broadcasting the Reactionary Right on YouTube." New York, NY. **https://datasociety.net/research/media-manipulation**.

Lorenz-Spreen, Philipp, Bjarke Mørch Mønsted, Philipp Hövel, and Sune Lehmann. 2019. "Accelerating Dynamics of Collective Attention." *Nature Communications* 10 (1): 1759. https://web.archive.org/web/20190614124801/**https://www.nature.com/articles/s41467-019-09311-w**.

Luis, Adriel, Lawrence-Minh Bùi Davis, Nafisa Isa, Kālewa Correa, Jeanny Kim, Hana Maruyama, Clara Kim, Nathan Kawanishi, Emmanuel Mones, Desun Oka, Carlo Tuason, Lisa Sasaki, Andrea Kim Neighbors, Deloris Perry, and Emily Alvey. 2017. "Culture Lab Manifesto." Poetry Magazine. July 2017.

M+. 2018. "About M+." https://web.archive.org/web/20180309043353/**https://www.westkowloon.hk/en/mplus/about-m**

Madrigal, Alexis C. 2017. "The Weird Thing About Today's Internet." *The Atlantic*. http://web.archive.org/web/20170520025036/**https://www.theatlantic.com/technology/archive/2017/05/a-very-brief-history-of-the-last-10-years-in-technology/526767/**

Merler, Michele, Nalini Ratha, Rogerio S. Feris, and John R. Smith. 2019. "Diversity in Faces," January. https://web.archive.org/web/20190712193151/**https://arxiv.org/abs/1901.10436**.

Merritt, Elizabeth. 2019. "TrendsWatch 2019." Center for the Future of Museums, Washington, DC.

Mitchell, Amy, Jeffrey Gottfried, Sophia Fedeli, Galen Stocking, and Mason Walker. 2019. "Many Americans Say Made-Up News Is a Critical Problem That Needs To Be Fixed." Pew Research Center. https://web.archive.org/web/20190923155418/**https://www.journalism.org/2019/06/05/many-americans-say-made-up-news-is-a-critical-problem-that-needs-to-be-fixed/**.

Murawski, Mike. 2017. "Museums Are Not Neutral." Art Museum Teaching. https://web.archive.org/web/20190712193214/**https://artmuseumteaching.com/2017/08/31/museums-are-not-neutral/**.

Oldenburg, Ray. 1989. *The Great Good Place: Cafés, Coffee Shops, Community Centers, Beauty Parlors, General Stores, Bars, Hangouts, and How They Get You through the Day*, 1st ed. New York: Paragon House.

O'Reilly, Tim. 2005. "What Is Web 2.0? Design Patterns and Business Models for the Next Generation of Software." OReilly.Com. http://web.archive.org/web/20100508031108/**http://oreilly.com/pub/a/web2/archive/what-is-web-20.html**

Parry, Ross. 2013. "The End of the Beginning." *Museum Worlds* 1 (1): 24–39. https://doi.org/10.3167/armw.2013.010103.

Pavis, Mathilde, and Andrea Wallace. 2019. "Response to the 2018 Sarr-Savoy Report: Statement on Intellectual Property Rights and Open Access Relevant to the Digitization and Restitution of African Cultural Heritage and Associated Materials," March. http://web.archive.org/web/20190610002217/**https://zenodo.org/record/2620597**

Price, Kati, and Dafydd James. 2018. "Structuring for Digital Success: A Global Survey of How Museums and Other Cultural Organisations Resource, Fund, and Structure their Digital Teams and Activity." Proceedings of Museums and the Web 2018. http://web.archive.org/web/20181019220314/**https://mw18.mwconf.org/paper/structuring-for-digital-success-a-global-survey-of-how-museums-and-other-cultural-organisations-resource-fund-and-structure-their-digital-teams-and-activity/**

Praagh, Shauna Van. 2005. "Preface Navigating the Transsystemic: A Course Syllabus." https://www.mcgill.ca/centre-crepeau/files/centre-crepeau/vanPraagh_Navigating.pdf.

Poole, Nicholas. 2014. "Change." CODE | WORDS: Technology and Theory in the Museum—Medium. https://web.archive.org/web/20190712193259/**https://medium.com/code-words-technology-and-theory-in-the-museum/change-cc3b714ba2a4**.

Reilly, Maura. 2018. *Curatorial Activism: Towards an Ethics of Curating*, 1st ed. London: Thames & Hudson.

Rekdal, Per B. 2014. "Why a Book on Museums and Truth?" In *Museums and Truth*, edited by Annette B. Fromm, Per B. Rekdal, and Viv. Golding, 1st ed., ix–xxv. Newcastle upon Tyne: Cambridge Scholars Publisher.

Rideout, Victoria, and Vikki S. Katz. 2016. "Opportunity for All? Technology and Learning in Lower-Income Families." www.vikkikatz.com.

Rodney, Seph. 2015. "How Museum Visitors Became Consumers." CultureCom. https://web.archive.org/web/20190712193442/**https://culture-communication.fr/en/how-museum-visitors-became-consumers/**.

———. 2016. "The Evolution of the Museum Visit, from Privilege to Personalized Experience." Hyperallergic. http://web.archive.org/web/20180629102144/**https://hyperallergic.com/267096/the-evolution-of-the-museum-visit-from-privilege-to-personalized-experience/**

"Russia Used Social Media for Widespread Meddling in U.S. Politics: Reports." 2018. Reuters. https://web.archive.org/web/20190712193530/**https://www.reuters.com/article/us-usa-trump-russia-socialmedia/russia-used-social-media-for-widespread-meddling-in-u-s-politics-reports-idUSKBN1OG257**.

Samis, Peter, and Mimi Michaelson. 2017. *Creating the Visitor-centred Museum*. New York: Routledge, Taylor & Francis Group.

Sanderhoff, Merete. 2017. "Open Access Can Never Be Bad News." SMK Open. https://web.archive.org/web/20190712193610/**https://medium.com/smk-open/open-access-can-never-be-bad-news-d33336aad382**.

Sardar, Ziauddin. 2010. "Welcome to Postnormal Times." *Futures* 42 (5): 435–44. https://web.archive.org/web/20190712193649/**https://ziauddinsardar.com/articles/welcome-postnormal-times**.

Scheiber, Noam, and Mike Isaac. 2019. "Facebook Halts Ad Targeting Cited in Bias Complaints." NYTimes.Com, March 19, 2019. https://www.nytimes.com/2019/03/19/technology/facebook-discrimination-ads.html

Schonfeld, Roger, Mariët Westermann, and Liam Sweeney. 2015. "The Andrew W. Mellon Foundation Art Museum Staff Demographic Survey." https://web.archive.org/web/20190712193740/**https://mellon.org/programs/arts-and-cultural-heritage/art-history-conservation-museums/demographic-survey/**.

Sentance, Nathan "Mudyi." 2018. "Your Neutral Is Not Our Neutral." Archival Decolonist [-O-]. https://web.archive.org/web/20190712193826/**https://archivaldecolonist.com/2018/01/18/your-neutral-is-not-our-neutral/**.

Simon, Nina. 2019. "From Risk-taker to Spacemaker: Reflections on Leading Change." Museum 2.0. https://web.archive.org/web/20190713002328/**http://museumtwo.blogspot.com/2019/03/from-risk-taker-to-spacemaker.html**.

Smith, Koven. 2019. "At What Cost?" https://web.archive.org/web/20190714165344/**https://kovenjsmith.com/post/the-values-of-museum-technology/**.
Solon, Olivia. 2019. "Facial Recognition's 'Dirty Little Secret': Millions of Online Photos Scraped without Consent." NBC News. https://web.archive.org/web/20190712194056/**https://www.nbcnews.com/tech/internet/facial-recognition-s-dirty-little-secret-millions-online-photos-scraped-n981921**.
Sønderskov, Kim Mannemar, and Peter Thisted Dinesen. 2016. "Trusting the State, Trusting Each Other? The Effect of Institutional Trust on Social Trust." *Political Behavior* 38 (1): 179–202. https://web.archive.org/web/20161203201250/**http://link.springer.com/article/10.1007/s11109-015-9322-8**.
Songlines: Tracking the Seven Sisters. National Museum of Australia, Canberra. 15 September 2017 to 28 February 2018 https://web.archive.org/web/20190712194413/**https://www.nma.gov.au/exhibitions/songlines**
Tallon, Loic. 2017. "Digital Is More Than a Department, It Is a Collective Responsibility." The Met Blog. October. https://web.archive.org/web/20190712194145/**https://www.metmuseum.org/blogs/now-at-the-met/2017/digital-future-at-the-met**.
Thomas, Hank Willis. 2017. "As said in round table discussion with SFMOMA staff during the museum's 2017/2018 strategic planning workshops where he was an invited participant." San Francisco. August 2017.
Topaz, Chad M., Bernhard Klingenberg, Daniel Turek, Brianna Heggeseth, Pamela E. Harris, Julie C. Blackwood, C. Ondine Chavoya, Steven Nelson, and Kevin M. Murphy. n.d. "Diversity of Artists in Major U.S. Museums." Accessed June 11, 2019. https://arxiv.org/pdf/1812.03899.pdf.
"Transsystemic Legal Education | Paul-André Crépeau Centre for Private and Comparative Law – McGill University." n.d. McGill.CA. Accessed March 27, 2019. https://web.archive.org/web/20190714165146/**https://www.mcgill.ca/centre-crepeau/projects/transsystemic**.
V&A Research. 2018. "Immersive Dickens: Prototyping a Mixed Reality Immersive Display of a Charles Dickens Manuscript." V&A Research Projects. https://web.archive.org/web/20190712194305/**https://www.vam.ac.uk/research/projects/immersive-dickens**.
Vira, Udit. 2019. "A Field Guide to the Living Internet · Nesta." Finding Ctrl. https://web.archive.org/web/20190606035455/**https://findingctrl.nesta.org.uk/field-guide-to-the-living-internet/**.
"Vision — OF/BY/FOR ALL." 2018. OF/BY/FOR ALL. 2018. https://www.ofbyforall.org/vision.
Wambold, Sarah, Marty Spellerberg. 2016. "Falk Meets Online Motivation: Results from a Nationwide Survey Project." Museums and Web. https://mw2016.museumsandtheweb.com/proposal/falk-meets-online-motivation-results-from-a-nationwide-survey-project/.
Ward, Esme. 2018. "Introducing Age-Friendly Culture." Manchester Museum, University of Manchester. Accessed June 12, 2019. https://documents.manchester.ac.uk/display.aspx?DocID=31582
Weil, Stephen E. 1999. "From Being about Something to Being for Somebody: The Ongoing Transformation of the American Museum on JSTOR." *Daedalus* 128 (3): 229–258.
Westermann, Mariët, Roger Schonfeld, and Liam Sweeney. 2019. "Art Museum Staff Demographic Survey 2018." https://web.archive.org/web/20190711212646/**https://sr.ithaka.org/publications/interrogating-institutional-practices-in-equity-diversity-and-inclusion/**.

Winesmith, Keir. 2017. "Against Linked Open Data." Medium.com. http://web.archive.org/web/20170630234622/**https://medium.com/@drkeir/against-linked-open-data-502a53b62fb**

Zuboff, Shoshana. 2015. "Big Other: Surveillance Capitalism and the Prospects of an Information Civilization." *Journal of Information Technology* 30 (1): 75–89. https://web.archive.org/web/20190331030357/**https://journals.sagepub.com/doi/10.1057/jit.2015.5**.

ABOUT THE AUTHORS

Prof. Keir Winesmith is Professor in Art & Design at the University of New South Wales (UNSW) and a digital strategy consultant to museums, galleries and universities. Keir holds a Media Arts Ph.D. from UNSW and BSc (hons) from the University of Sydney. Previously, he was CTO at *Old Ways, New*, Director of Digital Experience at the *San Francisco Museum of Modern Art*, and led the digital department at the *Museum of Contemporary Art* in Sydney. In 2018 he was named in *Fast Company*'s 100 Most Creative People in Business. Keir has given keynote and plenary presentations at many conferences including SXSW, National Digital Forum in New Zealand, Australian Museums & Galleries Association, MuseumNext, REMIX, Museums and the Web, Communicating the Arts, and the annual Museum Computer Network conference, where he serves on the board of directors. He has appeared in the *New York Times*, *San Francisco Chronicle*, *WIRED Magazine*, *National Public Radio*, *the Australian Broadcasting Corporation*, *Radio New Zealand* and the *BBC*. Keir lives in Sydney with Susan and their children, and is an avid (some might say obsessive) museum-goer.

Dr. Suse Anderson is Assistant Professor, Museum Studies at The George Washington University and host of *Museopunks* – the podcast for the progressive museum. Prior to joining GWU, Anderson was Director of Audience Experience at *The Baltimore Museum of Art*, where she was responsible for creating a seamless and positive experience of the *BMA* for visitors to the museum and across its digital platforms. In 2018/19, Anderson was a Visiting Technologist at the *Pew Center for Arts and Heritage* in Philadelphia, and served on Congresswoman Eleanor Norton's Congressional Art Competition Committee. She was recently President of MCN (Museum Computer Network) (2017–18) and was Program Co-Chair for its annual conference (2015–16). She has given keynote or plenary presentations at conferences including American Alliance of Museums

(AAM), Museums & the Web (U.S.), Museums Australia (AUS), INTERCOM, Small Museums Association, the Visitor Experience Conference and MCN. Anderson has also co-edited two experimental publishing efforts focussed on responsive discourse in the museum sector: *Humanizing the Digital: Unproceedings from the MCN 2018 conference* and *CODE | WORDS: Technology and theory in the museum*. Anderson holds a PhD (Fine Arts) from The University of Newcastle, Australia. Since moving to Baltimore in 2014, Anderson has fallen in love with the city she now calls home. You should visit her there sometime.

INDEX

"activation" 62
activism 85
admission: free 126–7; pricing models 126
affordances 214
agency 39
agile digital product development 107, 108, 219
Ahhhcade 75
AI Now Institute 14
algorithms 14, 16, 21, 88; deepfakes 17; recommendation engines 17
APIs 87, 132, 133, 183
Arab Spring 13, 179
archiving 50, 55, 103; co-developed content 46–7; community 45, 46, 47–8, 53, 55–6, 57, 58, 59; community-centred approach 3; digital preservation 137
art museums: DMA Friends program 35–6; in Vietnam 82; visitors 37
art teaching 67
art writing 66–7
Arts Council England 176
Asia 82; and complex geographies 66; performance art 65–6; *see also* Hong Kong
Asia Art Archive (AAA) 3, 61, 62, 64; multilingualism 68–9
audience evaluation 5
audiences 71, 88, 124, 177, 181, 211; affordances 214; attracting new 126, 128; beliefs 167–8; and big data 143, 144; and culture 174–5; demographic profiling 144, 147–8; digital 152; engagement 146; evaluation tools 142–3, 146–7; expression segment 149, 150; formative evaluations 150–1; forums 154; hard to reach 216–17; insights 143–4; listening to 153, 164; M+ 63; motivation-centred approach to 145; online 116; participation 103; political awareness 170; psychographic segmentation 143; segmentation 143, 144, 145, 147, 152, 156; sense of belonging 200–1; summative evaluations 150–1; surveys 143; targeting 149–50; threshold fear 177–8, 180, 187; time wallets 126, 128; at the *Victoria and Albert Museum* 108–9, 115; *see also* engagement; participation
augmented reality (AR) 87; Dunhuang caves project 97
Austin History Center 44, 45, 47, 55, 57
Australasian Association for Digital Humanities 96
Australian Centre for the Moving Image (ACMI): Chief Experience Officer 123; cinemas 127; exhibit curators 124; film festivals 127; professional development plan 129
authority 41, 76, 175; leadership 176; and open access 104
authorship 6, 13; shared 208–10, 218

Bahram, Sina 221
Balkanisation of culture 38, 39, 41

Barnes, The: Chief Experience Officer 123; *Let's Connect Philly* 125–6
Bass, Michael Thomas 190
Beaudoin, Paul 133
behaviour modification empires 214
Benezra, Neal 125
Berger, John, *Way of Seeing* 213
biases: algorithms 14; in online collections 13
Big Bang Data 135
Big Data 16, 143, 152–3
blogs 129–30; *Hyperallergic* 30, 35
Bourdieu, Pierre 127
boyd, danah 17
Brain Scoop 168, 169
Brand, Stewart, *The Whole Earth Catalog* 79
branding 131–2
broadcasting and journalism 5, 159, 165; museum partnerships 168
Brother Cane 65–6
Bruguera, Tania 41
Butt, Zoe 82

calls to action 161
canonisation 91
Cantu, Rossana 7
censorship 66; in China 82
Chicago, *Field Museum* 5, 159
Chief Experience Officers 123, 124
China, Dunhuang caves project 96–7
Chinese-language social media 69–70
classification 20
climate change 11
collaboration 102, 117, 129, 137, 186, 198, 218; with libraries 137–8; shared authorship 208–10
collections 40, 57, 119; co-developed 58–9; co-developed content 46–7; digitised 94, 95; encoding metadata 59; open access 100, 101, 103–4; *see also* online collections
Collins, Thom 124
colonialism, "adjustment of colonial terminology" project 99
community archiving 3, 45, 46, 47–8, 53, 55–6, 57, 58, 59
community service providers 148
community values 78
complex geographies 66
complexity 10, 11, 12, 19, 20
contemporary museums 63–4
contradictions 17, 20, 21
control of data 39
controversies 38–9

conversations, editing process 6–7
Cope, Aaron 16
copyright 71, 133
Cousins, Jill 221
Crawford, Kate 16–17
Crenshaw, Kimberlé 77
CrossLines 78
crowdfunding 185
cultural capital 203–4
cultural participation 71–2
cultural permissioning 57–8
culture 174, 175, 177, 179, 181; Balkanisation of 38, 39, 41
Culture is Digital report 5, 173, 180, 185
Culture Lab Manifesto 77–8
Culture Labs 3, 75, 76–7, 84; *CrossLines* 78, 80
Culture Track 173, 174, 176, 187, 202; *see also* audiences
curation 30, 40, 91, 92, 156, 174, 194, 211, 213; and collaboration 103; and community values 78; exhibit 124
curiosity 108, 161, 162, 223
cyberspace 12

Dallas Museum of Art, DMA Friends program 35, 36
data breaches 134–5
data silos 117
Davis, Lawrence-Minh Bùi 77
decision-making 209
deep data 143
deepfakes 17
demographic profiling 144; expression segment 149–50
Denmark: Dokk1 40; *IT-Universitetet i København* 3; SMK 91, 95–6
Derby Museums Trust 5, 190–1, 194; business plan 193–5; funding models 196, 197; *Museum of Making* 193, 195, 202; Silk Mill 193, 199
designing exhibitions 111–12
dialogues 1, 2; informal conversations 2
Dickens, Charles 118
"digital" 2, 7, 8, 62, 65, 72, 92, 182; APIs 87; at M+ 63; *see also* technology
digital calls to action 161
digital co-production 220
digital cultural heritage 97–8, 103
Digital Culture 180
"digital dilemma" 181, 187
digital museology 92
digital museum practice 92
digital practice 106, 107, 109, 110, 111, 112, 114, 120, 139, 180–1, 219, 222;

centralisation 110; and change 113; investing in 114; and transparency 112
digital preservation 133–4, 137
digitised collections 94, 95
Dilenschneider, Colleen 35
discrimination, algorithmic 14
Diversity in Faces (DiF) 14
DMA Friends program 35, 36, 37
docents, security guards as 35, 36
Doctorow, Cory 18
Does Art Have Users? 85
Dokk1 40
DomeLab 93
donations 196, 197
Dunhuang caves project 96–7

ECHO Lake Aquarium and Science Center 147
École Polytechnique Fédérale de Lausanne 4
engagement 146, 151–2, 159, 162, 163, 165, 177, 186, 198, 201, 204, 210; audience forums 154; communal experiences 161; and the "digital dilemma" 181; "ladder of" 178
epistemology, crisis in 18
Europeana 92, 96, 120, 221; Impact Framework 98
evaluation 151, 155–6, 203; and big data 152–3; conversations as 153; forums 154; NPS (Net Promoter Score) 202; *see also* audiences
Evolving Planet 167
exhibit curators 124
Exhibition Tax Relief 196
exhibitions 19–20; *Ahhhcade* 75; *Big Bang Data* 135; controversial 38–9; designing 111–12; *Evolving Planet* 167; *Invisible Intersections* 52; minimum viable product (MVP) 118; planning 78–9; *Songlines: Tracking the Seven Sisters* 93; *Tattoo* 162–3; *see also* Culture Lab
Expanded Perception and Interaction Centre (EPICentre) 91
expertise 76, 77, 96, 180, 186, 187, 213; and culture 175; and leadership 176; and neutrality 165; *see also* authority
expression segment 149, 150

Facebook 14, 17, 55, 84
facial recognition technology 14, 135
fake news 170
Falk, John 146, 147
Field Museum 159; *Brain Scoop* 168, 169
Flickr 132
formative evaluations 150–1

Formstack 132–3
Fox, Hannah 199
Fraser, John 219
free-entry institutions 145; *see also* admission
Fukuoka, Masanobu, *The One Straw Revolution* 79
funding models 138, 182, 184–5, 186, 196, 221; grants 176; micro grants 181
Funtowicz, Silvio 10

galleries: free 126–7; in Vietnam 82
galleries, libraries, archives and museums (GLAM) 1, 51, 53
games 64, 83
Gangsei, Erica 145
Gardner, Eric 221
General Data Protection Regulation (GPDR) 135, 136
Gibson, James J. 214
GIFT project 3, 75, 85
Gil, Britney 18
Global Financial Crisis (GFC) 197
globalisation 11
Goggin, Benjamin 14
Google 17
Google Arts and Culture Institute 14
Gottschall, Jonathan, *Storytelling Animal* 169
grants 94; micro 181; *see also* funding models
Graslie, Emily 168
Green, Hank 168
Grinell, Klas 20

Happy Museum Project 190, 194, 201
hard to reach audiences 216–17
Hernandez, Sarita 221
Holiday, J. F. 191
Hong Kong: *Asia Art Archive (AAA)* 3, 61, 62, 64; M+ 3, 61, 62, 63, 65, 66; political context 65, 66, 72; teaching art in 67; and video games 64; WeChat 70; *see also* Asia Art Archive (AAA); M+
human-centred design toolkit 199
Hyperallergic 30, 35

identity 61, 82, 146; and mission statements 85
images, usability 101; *see also* archiving
Immersive Dickens project 118
immersive experiences 102, 118–19, 120
Iñárritu, Alejandro, *Carne Y Arena* 120
inclusivity 6, 21, 177, 218, 221
Indego Bike Share 137
Indianapolis Museum of Art, IMA Lab 30
Indigenous knowledge 49, 50, 59

inequality, digital 13
informal conversations 2
Instagram 86
institutional bias 13
institutional change 84, 85, 86
institutional trust 18, 164–5, 166–7
interdisciplinarity 87
International Image Interoperability Framework (IIIF) 101
internet, the 11, 33, 54, 91, 99, 190; and authorship 13; online audiences 116; online collections 13, 14
intersectionality 77, 78
investing in digital 114
IT-Universitetet i København 3

James, Dafydd 106
Johnston, Courtney 139, 223

kaitiaki 47
Kane, Natalie 15
Kearney, Caitlin 164
Kiribati magic 50
knowledge 11, 12, 17, 20, 34, 76, 79, 93, 118, 213, 220; crisis in epistemology 18; Indigenous 49, 50, 59; sacred 50
koha 52
Kuhn, Thomas 10

Laboratory for Innovation in Galleries, Libraries, Archives and Museums (iGLAM) 91
labs 183, 184
Lam, Carrie 66
Langenbach, Ray 65–6
LaPlaca Cohen 173; "Culture Track" report 5
Latour, Bruno 97
leadership 176
learning 32, 38, 184, 216
libraries 41, 80, 137–8, 144; Dokk1 40; *see also* collaboration
licensing 100–1
listening 164
London: Science Gallery London 5; *South Bank Centre* 179; Victoria and Albert Museum 4
London, Asia 66
Lots of Copies, Keeps Stuff Safe (LOCKSS) 53–4
Lucas Museum 40

M+ 3, 61, 62, 63, 65, 66; audience 63; multiculturalism 68; *Podium* 68; Sigg Collection 65

M+ Matters 68
M+ Stories website 68
machine learning 17
mailing lists 136
marketing 31, 39, 70, 110, 131, 155, 198
McGregor, Liz-Anne 35
McKinley, Kelly 202
Medium 4, 123
MEOW WOLF 75
Met Open Access project 100–1
Metropolitan Museum of Art 106; audience 115; digital department 107; Met Open Access project 100–1; online audience 116; open access policy 118; stakeholder management 107–8
Michaelson, Mimi, *Creating the Visitor-Centred Museum* 39
micro grants 181
minimum viable product (MVP) 118
mission statements 85
modernisation 11
Mollica, Jay 183
Morris Hargreaves McIntyre (MHM) 142; audience evaluation tools 142–3
Mukurtu CMS 59
multiculturalism, *M+* 68
multilingualism, *Asia Art Archive (AAA)* 68–9
Museum of Australian Democracy (MoAD) 5, 160, 168, 169
Museum of Making 193, 195, 196, 202
museums 12, 21; authority structure 33, 41; Chief Experience Officers 124; collection systems 20; contemporary 63–4; and controversy 38–9; data breaches 134–5; digital practitioners 88; exhibitions 19–20; free-entry 145; funding models 138; grants 94, 176; labs 183, 184; mailing lists 136; marketing 31; mission statements 85; neutrality of 19, 33, 165, 166, 167; normal practice 11; outreach 86–7; philanthropic models 136–7; pricing models 126; responsiveness 79–80; staff salaries 37–8; stakeholder management 84; Starbucks model 178; tools 39; and trust 18–19; trust in 34, 41; TV model 160; use of surveillance technologies 14–15, 16; "useful" 38, 41; virtual 3; visitor-centered 32; *see also* collections; exhibitions
music 217

National Gallery of Denmark (SMK) 91, 92; digital strategy 95–6
Neighbors, Andrea Kim 77

Nesta report 138–9
networks 11, 13, 16, 21
New Orleans, Cultural Heritage & Social Change Summit 3
New York, *Metropolitan Museum of Art* 4
New Zealand: Pacific Collections Access Project (PCAP) 43, 48–9, 55, 56; *Tamaki Paenga Hira Auckland War Memorial Museum* 3, 43
normal practice 11
NPS (Net Promoter Score) 202

Oakland Museum of California 191; funding 195, 197; organisational structure 197–8
online audiences 116
online collections 14, 21, 55; barriers to access 13; scalability 119
open access 13, 100, 101, 103–4; at the *Metropolitan Museum of Art* 118
outsourcing websites 111
Oxford, Trish 221

Pacific Collections Access Project (PCAP) 43, 46, 47, 52, 53, 55, 56; *teu le va* 48–9
Parry, Ross 219
participation 6, 20, 103, 178, 194, 203, 211; community 52; *Let's Connect Philly* 125–6; threshold fear 177–8, 180
partnerships 168–9, 170, 199; *see also* collaboration
Patton, Michael Quinn 151
performance art 62, 65–6
permaculture 79
personalisation 3, 29, 30–1, 34, 35, 39, 41
personas 145
philanthropic models 30, 149, 184–5, 186, 195, 197
Pinterest 117
planning exhibitions 78–9
play 6, 212
playfulness 83
PlaySFMOMA Mixed Reality Pop Up Arcade 75
Poole, Nicolas 11
"post normal science" 10
post-custodial model 53–4
postnormal times 20, 21; complexity 10, 11; contradictions 12, 17; surveillance capitalism 16
prestige transfer 83, 149
privacy 39, 134, 135; General Data Protection Regulation (GPDR) 135, 136; mailing lists 136; surveillance capitalism 16–17
product culture 108, 110, 111

protests, against exhibitions 38
protocols 59
psychographic segmentation 143, 145
public trust 80, 81; and fake news 170

Raicovich, Laura 21, 85
Rapid Response Collecting 115
Ravetz, Jerome 10
recommendation engines 17
reflective practice 139
relationships 8, 161, 193, 212, 213, 218; classification 20; community 52–3; *teu le va* 48–9
research 2
return on investment (ROI) 114, 116, 118
Rijksmuseum, "adjustment of colonial terminology" project 99
Roach, Kevin 191
Ross, Jen 220
Ross, Joan 127

sacred knowledge 50
salary of museum staff 37–8, 138
Samis, Peter, *Creating the Visitor-Centred Museum* 39
San Francisco Metropolitan Museum of Art (SFMOMA) 75, 153, 182, 183, 184, 200, 202; *Send Me SFMOMA* 136, 183
Sardar, Ziauddin 8, 10, 12, 17, 20
science, "post normal" 10
Science Gallery 207, 209, 213, 214, 216; themes 211
security guards, as docents 35, 36
segmentation 143, 144, 145, 147, 152, 156
Send Me SFMOMA 136, 183
shared authorship 208–10, 218
showcases 111; at the *Victoria and Albert Museum* 108–9
Sickle Cell Disease 215–16
Sigg, Uli 64
Silk Mill 199
Simon, Nina 220
Smith, Koven 222–3
Smithsonian Asian Pacific American Center 3, 75, 76, 78, 79, 80, 81, 82, 84, 86, 87
Snapchat 86
Snowden, Edward 115
social capital 127, 194
social change 84
social media 13, 17, 38, 54–5, 171; Chinese-language 69–70; collaborations 117; Facebook 14; Instagram 86; Pinterest 117; recommendation engines 17; Snapchat 86; WeChat 69, 70, 115; YouTube 217

social return on investment (SROI) 185–6, 201, 202
Songlines: Tracking the Seven Sisters 93
South Bank Centre 179
Stack, John 7
stakeholder management 84, 103; Metropolitan Museum of Art 107–8
Starbucks model 178
storytelling 114–15, 169, 208
summative evaluations 150–1
surveillance capitalism 16, 88; facial recognition technology 135
surveillance technologies 14–15, 16, 21, 222–3
sustainability 69, 190, 194; of skilled staff 138
Switzerland, École Polytechnique Fédérale de Lausanne 4

Taking Part 173
Tamaki Paenga Hira Auckland War Memorial Museum 3, 43
taonga 43, 44, 45, 46, 56
Tate Modern 34, 35
Tattoo 162–3
teamLab 206–8, 209, 211, 212, 214, 215
technology 88, 93, 94, 96, 163, 214, 221; adopting online platforms 130–2; APIs 87, 132, 133, 183; augmented reality (AR) 87, 97; facial recognition 135; fragility of 131, 132; virtual reality (VR) 87
teu le va 48, 53
Thomas, Hank Willis 34
Thompson, Christian 127
threshold fear 177–8, 180, 187
tikanga 50, 56
time wallets 126, 128
Tokyo, teamLab 5–6
tools 87, 200; *Culture Track* 173, 174, 176, 202; human-centred design toolkit 199
topic maps 8
Tough Blood 215–16
Traditional Knowledge (TK) Labels 59
translation 73, 77, 119; *see also* multilingualism
transparency 128
trust 41, 50, 78, 80, 82, 86, 166, 167, 171, 179, 184; in artists 80; and fake news 170; institutional 18, 164–5; in museums 18–19; public 80–1
truth 17, 18
TV model 160

"useful museums" 38, 41
user acceptance (UA) 151–2
user experience (UX) 151–2

values 78, 160
van Heeswijk, Jeanne, *Philadelphia Assembled* 51, 52
Victoria and Albert Museum 1, 114; audience 108–9, 115; digital department 108; Immersive Dickens project 118; Rapid Response Collecting 115; showcases 108–9
video games 64
virtual museums 3
virtual reality (VR) 87, 127
visit modes 147
visitor experiences 4, 30–1, 35, 36; affordances 214; evaluation 203; and surveillance technology 15–16; *see also* audiences; evaluation
Visser, Jasper 92
visual culture 63–4; *Podium* 68
volunteering 193

Walker, Kelley 38
Ward, Esme 201
websites, outsourcing 111
WeChat 69, 70, 115
Weil, Stephen 219
well-being 201
Whitaker, Meredith 14
Wikidata 117
WikiLabs 95
Wikipedia 94, 116, 117, 118
Wittman, Aletheia 221
Wong, Jen 211
Wright, Joseph 194

Yeung, Ricky 67
YouTube 17, 217

zoos 144
Zuboff, Shoshana 16

Printed in Great Britain
by Amazon